Carlo Testa   The Industrialization of Building

Carlo Testa

The Industrialization of Building

Van Nostrand Reinhold Company
New York   Cincinnati   Toronto   London   Melbourne

Van Nostrand Reinhold Company Regional Offices:
New York Cincinnati Chicago Millbrae Dallas

Van Nostrand Reinhold Company International Offices:
London Toronto Melbourne

Copyright © 1972 by Verlag für Architektur Artemis Zürich
Library of Congress Catalog Card Number 72-12751
ISBN 0-442-28457-8

All rights reserved. No part of this work covered by the copyright hereon may
be reproduced or used in any form or by any means–graphic, electronic,
or mechanical, including photocopying, recording, taping, or information storage
and retrieval systems–without written permission of the publisher.

Printed in Switzerland

Published by Van Nostrand Reinhold Company
A Division of Litton Educational Publishing, Inc.
450 West 33rd Street, New York, N.Y. 10001

Published simultaneously in Canada by
Van Nostrand Reinhold Ltd.

16 15 14 13 12 11 10 9 8 7 6 5 4 3 2 1

Inhaltsverzeichnis

Index

Vorwort		7
1.	Einführung	9
1.1	Eine gemeinsame Terminologie	9
1.2	Was unterscheidet industrialisiertes Bauen vom konventionellen Bauen?	20
2.	Einschätzung der heutigen Lage	23
2.1	Geschichtliche Entwicklung	23
2.2	Erfolg und Mißerfolg	24
2.3	Analyse einiger erfolgreicher Unternehmen	26
2.4	Die Vorteile des industrialisierten Bauens	32
2.5	Industrialisierung des Bauens und die Entwicklungsländer	34
3.	Architektur und Industrialisierung	37
3.1	Die Situation	37
3.2	Die Schwierigkeiten des ausgebildeten Architekten von heute	40
3.3	Die mögliche Stellung des Architekten	42
3.31	Forschung	42
3.32	Entwicklung	43
3.33	Projektleitung	45
3.34	Bauentwurf und Stadt- und Regionalplanung	47
3.35	Zusammenfassung	48
4.	Organisatorische Konzepte	49
4.1	Wer hat die Bauindustrialisierung gefördert?	49
4.2	Wie werden industrialisierte Unternehmen betrieben?	50
4.3	Organisatorische Schwierigkeiten	54
4.31	Der Entscheidungsprozeß	54
4.32	Beschaffung von Informationen	54
4.33	Verarbeitung von Informationen	56
4.34	Produktion	58
4.35	Montage	59
4.4	Operationelle Taktiken	60
4.5	Die Nebenindustrie	65
4.6	Mögliche Organisationsformen	65
5.	Konkrete Beispiele	67
5.1	Einleitung	67
5.2	Studienbeispiele	
	National Home Corporation, USA	70
	Delta Homes Corporation, USA	76

Foreword		7
1.	Introduction	9
1.1	A common terminology	9
1.2	What makes industrialized building different from conventional building?	20
2.	Assessment of the Present Situation	23
2.1	Historical development	23
2.2	Success and failure	24
2.3	An analysis of some successful ventures	26
2.4	The advantages of the industrialized building	32
2.5	Industrialization of building and the developping countries	34
3.	Architecture and Industrialization	37
3.1	The situation	37
3.2	The difficulties of the educated architect today	40
3.3	The possible position of the architect	42
3.31	Research	42
3.32	Development	43
3.33	Project management	45
3.34	Building design and urban and regional planning	47
3.35	Conclusions	48
4.	Organizational Concepts	49
4.1	Who has been behind the industrialization of building?	49
4.2	How have industrialized ventures operated?	50
4.3	Organizational difficulties	54
4.31	Decision making	54
4.32	Production of information	54
4.33	Processing of information	56
4.34	Production	58
4.35	Assembly	59
4.4	Operational tactics	60
4.5	The subsidiary industry	65
4.6	Possible organizational forms	65
5.	Case Studies	67
5.1	Introduction	67
5.2	Case studies	
	National Home Corporation, U.S.A.	70
	Delta Homes Corporation, U.S.A.	76
	IPI S.p.A., Italy	82

	IPI S.p.A., Italien	82
	Impresa Generale Costruzioni MBM Meregaglia S.p.A., Italien . . .	88
	Balency & Schuhl, Frankreich . .	94
	Constructions Modulaires S.A., Frankreich	100
	Carl Möller, Bundesrepublik Deutschland	106
	Nachbarschulte & Co. KG, Bundesrepublik Deutschland . .	112
	IGECO, Schweiz	118
	Vic Hallam Limited, England . .	124
	South Eastern Architect Collaboration (SEAC), England	130
	Kombinat für Wohnungsbau Nr. 3, UdSSR	136
	Kombinat für Wohnungsbau Nr. 1, UdSSR	142
	Servicio Técnico de Construcciones Modulares S.A., Spanien . . .	146
	Staatliche Baugesellschaft Nr. 43, Ungarn	152
5.3	Bemerkungen	158
6.	Meinungen der Experten	163
6.1	Einleitung	163
6.2	Expertenmeinungen Anthony Williams, England . . .	163
	Thomas Schmid, Schweiz . . .	164
	Masayumi Yokoyama, Schweiz . .	166
	Wolfgang Döring, Bundesrepublik Deutschland	167
	David Chasanow, UdSSR . . .	169
	Jean Prouvé, Frankreich	170
	Maurice Silvy, Frankreich . . .	171
	Riccardo Meregaglia, Italien . .	173
	Robertson Ward, USA	176
	Gyula Sebestyén, Ungarn . . .	177
7.	Zukünftige Entwicklungen . . .	179
7.1	Technologische und technische Entwicklungen	179
7.2	Organisatorische Entwicklungen . .	181
7.3	Architektonische Entwicklungen . .	184
8.	Schlußfolgerungen	190
8.1	Die Zukunft des industrialisierten Bauens	190
8.2	Die Notwendigkeit eines Kontrollorgans	192
9.	Verzeichnis der Begriffe	195
10.	Bibliographie	197
11.	Organisationen, die sich mit dem industrialisierten Bauen beschäftigen .	199

	Impresa Generale Costruzioni MBM Meregaglia S.p.A., Italy . . .	88
	Balency & Schuhl, France . . .	94
	Constructions Modulaires S.A., France .	100
	Carl Möller, German Federal Republic . .	106
	Nachbarschulte & Co. KG, German Federal Republic	112
	IGECO, Switzerland	118
	Vic Hallam Limited, England . .	124
	South Eastern Architect Collaboration (SEAC), England	130
	House-Building Combine No. 3, U.S.S.R.	136
	Integrated Home Building Factory No. 1, U.S.S.R.	142
	Servicio Técnico de Construcciones Modulares, Spain	146
	State Building Organization No. 43, Hungary	152
5.3	Remarks	158
6.	Opinions of Experts	163
6.1	Introduction	163
6.2	Opinions of experts Anthony Williams, England . . .	163
	Thomas Schmid, Switzerland . .	164
	Masayumi Yokoyama, Switzerland . .	166
	Wolfgang Döring, German Federal Republic . . .	167
	David Khazanov, U.S.S.R. . . .	169
	Jean Prouvé, France	170
	Maurice Silvy, France	171
	Riccardo Meregaglia, Italy . . .	173
	Robertson Ward, U.S.A.	176
	Gyula Sebestyén, Hungary . . .	177
7.	Future Developments	179
7.1	Technological and technical developments	179
7.2	Organizational developments . . .	181
7.3	Architectural developments . . .	184
8.	Conclusions	190
8.1	The future of industrialized building .	190
8.2	The need for a controlling authority .	192
9.	Glossary	195
10.	Bibliography	197
11.	Relevant Organizations Interested in the Industrialization of Building . . .	199

Vorwort

Das vorliegende Werk befaßt sich mit der Industrialisierung des Bauens von innen (von der täglichen Praxis und Erfahrung) her gesehen. Dabei habe ich mich besonders auf die Probleme der Fachleute, ihre Funktionen und den von ihnen gewählten organisatorischen Aufbau konzentriert. Auch habe ich versucht, die Wechselwirkung zwischen ihrer Tätigkeit und dem Partner der «Außenwelt» (politische Macht, Amtsstellen, Benutzer) darzustellen. Es ist nicht meine Absicht, technische Lösungen oder spezifische Technologien zu analysieren. Für solche Information wird auf die Werke anderer Autoren verwiesen, die in der Bibliographie aufgeführt sind.
Ich glaube, daß dieses Werk dem lernenden und dem ausübenden Baufachmann von Nutzen sein kann. Meines Wissens gibt es kein Werk, in dem die praktischen Aspekte des industrialisierten Bauens als ein vollständiges System behandelt werden. Ein Hauptaspekt dieses Themas, der Entscheidungsprozeß der Behörde, kann hier nicht ausführlich behandelt werden.
Man kann täglich beobachten, wie die Behörden in allen Ländern eine stetig wachsende Rolle im Wohnungsbau spielen und vor allem über menschliche Wohnverhältnisse entscheidend bestimmen. Vorschläge, die hier für die Kontrolle der Amtsgewalt gemacht werden, sind zweifellos unzureichend. Man weiß, daß die Behörden gewöhnlich nur dann einer unhaltbaren Situation Rechnung tragen, wenn der Druck der Öffentlichkeit so stark wird, daß dadurch Stellung und Autorität der Verantwortlichen gefährdet sind. Die Ghettoaufstände in den Vereinigten Staaten, die Kämpfe in Italien für Sozialwohnungen, die Revolte der französischen Studenten — sie alle sind Beispiele aus letzter Zeit. Die Schwierigkeit liegt darin, daß Regierungen selten mit Phantasie auf den Druck der Öffentlichkeit reagieren: die so erzwungene Lösung betrifft immer nur eine Aufgabe, die schon vor den öffentlichen Protesten bestand. Hinzu kommen unvermeidliche zeitliche Verzögerungen, bis die Pläne tatsächlich realisiert werden und damit meistens auch die Lösung überholt ist, bis die Bauten endlich zustandekommen. So baute man neue Satellitenstädte in der Umgebung von Paris, als die Einwohner begannen, die Großstadt zu hassen; man baute Schulen in vielen sizilianischen Dörfern, nachdem die Eltern mit den Kindern bereits in den italienischen Norden gezogen waren; endlich baute man Straßen in Gebirgsgegenden, deren Bewohner das Tal längst verlassen

Foreword

This work is concerned with the industrialization of building as seen from the inside. I have dwelled upon the problems of the actors in this activity, on their rôles, on the organizational structure they have adopted. I have also attempted to indicate how the "outside" (the political power, the bureaucratic structure, the user), interreacts with their activity. It has not been my objective to analyse technical solutions or specific technologies. For this kind of information references are made to the works of other authors, indicated in the bibliography.
I feel that this work can be useful to the student of the building trade and to the practitioner of the building activity. To my knowledge there is no other work which considers the operational aspects of the industrialization of building as a complete system, as I have attempted to do in this book. However, a major facet of the subject has not been fully covered: the decision-making process proper to the public power.
We see daily how in all countries, the public power plays an increasingly important rôle in the construction of dwellings and generally in the determination of the conditions of the human habitat. The few proposals I have made in this work for the control of the power of the public authority are certainly insufficient. We know how the public power usually reacts to a situation only when the public pressure grows to the point of endangering the position of the people in charge (the revolt of the ghettos in the United States, the strikes in Italy to obtain low cost housing, the revolt of the students in France, are all recent examples). The trouble is that seldom do governments react to public pressure with a certain amount of imagination: the solution offered is always one, which tends to satisfy a requirement already existent before the public outburst occurred. Adding to this the unavoidable delays caused by the time required to transform plans into reality, we usually find that the solution is obsolete when the buildings are finally inaugurated. The new towns around Paris, built when people have come to hate the cities, the schools of too many Sicilian villages, built when the children with their parents have moved north, the roads in some mountain areas, built when the populations has long since abandoned the valley — these and hundreds more are living proofs of this disastrous time-gap.
If we assume that the industrialization of building, as exposed in this work, is a viable and satisfactory tool, we have to consider who should utilize this tool and

hatten — Beispiele, die mit hundert anderen lebendiges Zeugnis für solch unheilvolle Verzögerungen sind. Wenn wir annehmen, daß die hier beschriebene Industrialisierung des Bauens ein lebensfähiges und zur Lösung der Probleme wohlangemessenes Instrument ist, dann muß folgerichtig überlegt werden, von wem und wie dieses Instrument angewandt werden kann. Die Behandlung dieses Themas soll einer besonderen Darstellung vorbehalten bleiben.

Der Leser wird feststellen, daß Widersprüchlichkeiten dem Thema entsprechen und nicht zu vermeiden waren. Die Industrialisierung des Bauens ist immer noch ein junges Phänomen, und viele Meinungen und Auffassungen können und müssen zum Vorteil der Entwicklung modifiziert werden. Der Kampf zwischen Architekten, Industriellen und Bauunternehmern um die Führung ist gesund und verständlich. Auf Grund meiner eigenen Kenntnisse und Erfahrungen bin ich oft versucht, den Führungsanspruch der einen oder anderen Gruppe zu unterstützen. Denn wir werden in den nächsten Jahren höchstwahrscheinlich erleben, daß die führende Rolle nicht von einer bestimmten traditionellen Berufsgruppe übernommen wird, sondern von Einzelnen oder von Gruppen, die sich auf dem *gesamten* Wirkungsfeld der Bauindustrialisierung auskennen.

Zuletzt eine Warnung: die Unternehmungen, die ich ausgewählt, und die Produkte, die ich genannt habe, sollen den heutigen Verhältnissen realistisch entsprechen. Besonders habe ich vermieden, jene seltenen Beispiele «guter» Architektur anzuführen, welche für unsere Gegenwart ohne Bedeutung sind. Die vorliegende Arbeit will den heutigen Stand der Technik darstellen. Meine Absicht ist, dem Leser eine Basis anzubieten: von dort gibt es verschiedene Entwicklungsmöglichkeiten.

how. Another work will throw some light on this subject.

In my work I have not attempted to avoid contradictions. The industrialization of building is still a recent phenomenon, and many positions and attitudes can still be modified advantageously. The fight for the leading rôle between architects, industrialists, builders, is sound and acceptable. Due to my own background I often feel tempted to support the one or the other in their claims for leadership. Most likely we shall see in the next years how the leadership rôle will be taken not by a specific profession, but by individuals or groups emerging from any one of the actors of the *entire* field.

A last warning: The organizations I have selected and the products shown have been chosen to represent the reality of to-day. I have specially avoided giving rarified examples of "good" design which have no significance in the general picture. This work attempts to demonstrate the state of the art of today. It is my intention to provide the reader with a starting-point: from here, all kinds of developments may take place.

1. Introduction

1.1 A common terminology

It is necessary, in order to carry on a meaningful conversation, that all parties agree on the definition of certain terms. I would have liked to choose these few definitions from established sources (and for most of them such as module, joint, etc. clear definitions are indeed available, see "Glossary"). But for the more general terms (like industrialization, prefabrication, etc.) it seems that a considerable confusion exists. I have therefore to propose a few definitions which will have to be accepted at least as a working hypothesis within the context of this book. To begin with, what do we understand by "the industrialization of building"? I have found a few definitions by various authors and organizations, but none of them seems to be accepted worldwidely and they do not seem to cover the full meaning of the term. The definition I propose is: industrialization is a process which, by means of technological developments, organizational concepts and methods, and capital investment, tends to increase productivity and to upgrade performance. The advantage of this definition is that it covers the field of building and that it allows us to break down a general concept into clearly differentiated forms of applications.

We also dispose of two objectives, productivity and performance, which can be transformed into parameters against which we can measure the success of our activity. It also seems to provide a basis (see 1.2) for distinguishing industrialized building- from conventional building-methods.

If we accept the above definition as a working hypothesis we can now identify several forms of industrialization currently used in the building trade. A first analysis shows four forms of industrialization used in the world at present:
a) Prefabrication
b) Modular system building
c) Rationalized building
d) Equipment-oriented site-production

Before we define and exemplify these four forms of industrialization, it must first be understood that the above list is not exclusive and new forms may and will develop, and second that the above forms are not necessarily mutually exclusive but that they can and do coexist on the same building-project or within the same organization.

The definition of *prefabrication* is contained in the

Bei der Vorfertigung wird zunächst das Gebäude entworfen, dann wird es für die Produktion in sinnvolle Teile unterteilt, und schließlich werden die Teile auf der Baustelle montiert.

In prefabrication you first design the building and then you decompose it in meaningful parts for production and finally the parts are assembled on site.

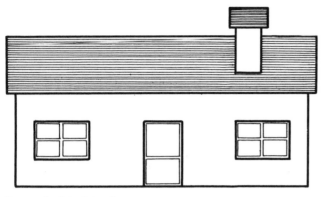

Der ursprüngliche Entwurf.

The original design.

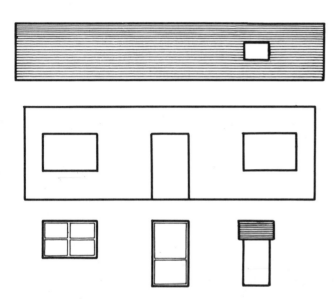

Das Bauwerk wird in seine Komponenten zerlegt, die dann in der Fabrik vorgefertigt werden.

The building is decomposed in its components and then the components are factory produced.

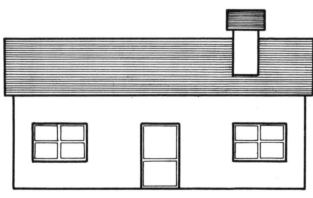

Das montierte Bauwerk.

The reassembled building.

▶

Terrapin Ltd., England. Ein sehr fortschrittliches Beispiel der Vorfertigung. Vollständige Sektionen eines Gebäudes werden in der Fabrik gefertigt, zur Baustelle transportiert und in wenigen Stunden montiert.

Terrapin Ltd., England. A very sophisticated case of prefabrication. Complete sections of a building are produced in the factory, transported to the site and assembled in a few hours.

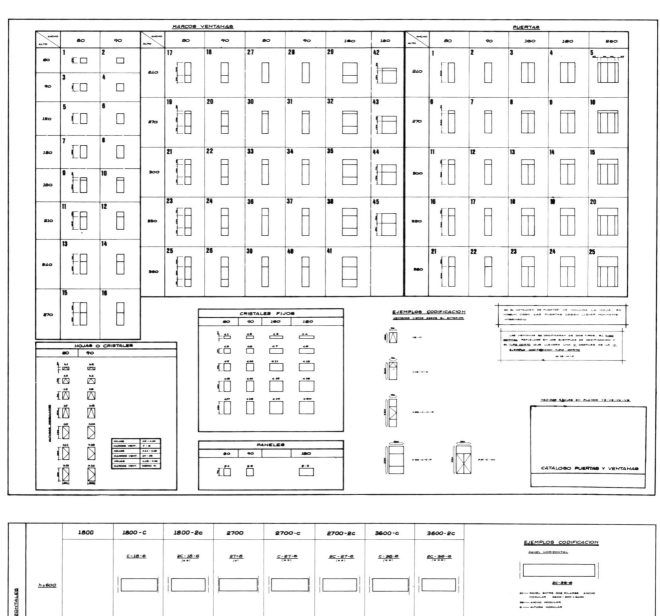
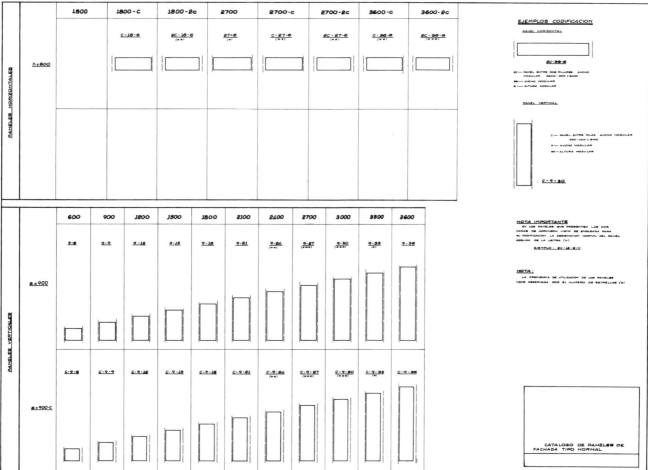

Servicio Técnico de Construcciones Modulares, S.A., Spanien. Katalog der Komponenten für die Subsysteme Außenwand (Fenster und Opakwand).

Servicio Técnico de Construcciones Modulares, S.A., Spain. The catalogues of components for the outside envelope sub-systems (windows and opaque wall).

Bevor wir diese vier Arten der Industrialisierung definieren und durch Beispiele erläutern, muß erstens klargemacht werden, daß die Liste neue Arten, die sich entwickeln können und werden, nicht ausschließt. Zweitens schließen sich die Methoden nicht notwendigerweise gegenseitig aus, sondern können in demselben Bauprojekt oder innerhalb der gleichen Bauunternehmung ergänzend angewandt werden.

Die Definition des Begriffs «Vorfertigung» ist durch das Wort selbst gegeben. Diese Art der Industrialisierung besteht aus der Herstellung von Teilen, die zusammengefügt ein fertiges Produkt ergeben. Ob diese Teile in einer Fabrik oder an der Baustelle hergestellt werden, ändert nicht das Wesentliche dieser Methode. Von grundsätzlicher Bedeutung ist die Tatsache, daß man bei der Anwendung dieser Industrialisierungsmethode zuerst das Fertigungsprodukt — sei es ein Bau oder zum Beispiel ein Gewehr — plant, daß man es dann in sinngemäße Teile — vom Standpunkt der Produktion und Montage — aufteilen, diese Teile herstellen und schließlich in der richtigen Reihenfolge und Ordnung zusammensetzen muß. Logischerweise erhält man immer das gleiche Produkt, wenn man die Teile zusammenfügt. Das kann leicht bewiesen werden, indem man den Vergaser eines Autos in seine Teile zerlegt und wieder zusammensetzt. Die Möglichkeit besteht natürlich, daß man die Teile in einer anderen Reihenfolge zusammenfügt — dann ergibt sich vielleicht eine moderne Plastik an Stelle eines Vergasers — oder daß man einige Teile so konstruiert, daß bei Hinzufügung einiger zusätzlicher Komponenten ein ganz neues Resultat erzielt werden kann. Dieses trifft — darum dieses Beispiel — auf einige Schußwaffen zu, bei denen der Abzug, die Trommel usw. und ein verlängerter Lauf die Leistung verändern.

Grundsätzlich betrachten wir jedoch die Vorfertigung als eine Art der Industrialisierung, bei der die Teile immer das vorhergeplante Endprodukt ergeben.

«Modulare Bausysteme» sind in ihrem intellektuellen Vorgang nach der Vorfertigung genau entgegengesetzt. In diesem Ablauf wird zunächst ein Satz von dimensional und funktionell zusammengehörigen Komponenten geplant, dann werden allgemeine Regeln festgesetzt, wie diese Teile miteinander verbunden werden können, und das Endprodukt wird dann aus diesen Komponenten konstruiert. Das einfachste Beispiel eines modularen Bausystems ist das «Meccano»-Spielzeug. Ein Satz von vorher bestimmten Teilen kann hier nach gewissen dimensionalen Grundregeln in eine unendliche Vielzahl von Gestaltungen zusammengesetzt werden. Im Baugewerbe müssen die Entwerfer immer mit der Festsetzung gewisser Hypothesen der Form und Dimension beginnen. Trotzdem aber hat man bewiesen, daß mit einem geeigneten «Katalog» der Komponenten eine erhebliche Flexibilität erreicht werden kann. Die Komponenten selbst können in einer oder in mehreren Fabriken hergestellt werden; aber man könnte sie kaum an der Baustelle fertigen, ohne die Begriffsbestimmung dieser Art Industrialisierung aufzugeben. Die Fabrikation der Teile an Ort würde bedeuten, daß sie für ein bestimmtes Bauwerk speziell hergestellt sind und damit als «Vorfertigung» zu bezeichnen wären.

Wenn dimensionale Ordnungen in einem Land oder zwischen mehreren Organisationen und Unternehmungen festgesetzt werden und eine gemeinsam

word itself. This form of industrialization consists of producing parts which, when assembled, will give a finished product. Whether these parts are produced in a factory or under the open sky does not change the essence of the method (factory-prefabrication and site-prefabrication). More relevant is the fact that when utilizing this form of industrialization, we have first to design the finished product (be it a building or a gun), then break it down into meaningful parts (from a production and assembly viewpoint), produce these parts, and finally assemble them in the correct sequence and order. Logically we shall always obtain the same product if we assemble the parts, as can easily be proved by taking a carburettor of a car, dismantle it in its parts and reassemble it again. There still exists a possibility that we assemble the parts in a different order (and then we may obtain a sculpture instead of a carburettor) or that we have designed some parts in such a way that with some additional parts a new performance may be achieved. (As is the case with some guns, where the trigger, the drum, etc., plus a longer barrel will alter the performance.) Basically however we shall look upon prefabrication as a form of industrialization where the parts will always give the same predesigned product.

Modular system building operate according to the opposite conceptual process of prefabrication. Here we first design a set of dimensionally and functionally inter-related components, we establish general rules of how these components may be connected together and with these components we then design a product. The simplest example of a modular system is the toy called "meccano". Here a set of predesigned parts can be assembled following certain basic dimensional rules to obtain an infinite variety of products. In practice in the building trade the designers of the components will always have to start by establishing certain hypothesis of forms and dimensions, but not withstanding this, it has been demonstrated that with a suitable "catalogue" of components, considerable design flexibility can be achieved. The components again can be produced in a factory or in several factories, but can hardly be produced on site without abandoning the conceptual basis of this form of industrialization. (Should we produce the parts on site we would be producing parts for a specific building and we would therefore fall back on "prefabrication".) Should a dimensional discipline be accepted in a given country or between several organizations and a predefined jointing technique be added, we would create a more sophisticated form of modular system which is defined as "componenting". While a modular building-system only allows for certain parts described in a catalogue to be utilized by the architect, in "componenting" we extend the freedom of the design, allowing him to choose from all kind of components produced. Basically the "meccano" becomes bigger and the components are made available by several independent sources.

Universität York, England; Robert Mathew, Johnson und Marshall, Architekten. Beispiel eines Gebäudes, das aus einem bestehenden Satz von Komponenten hergestellt (CLASP modulares Bausystem) und von einem qualifizierten Architekten entworfen ist.

York University, England, Robert Mathew, Johnson and Marshall, Architects. An example of a building produced from an established set of components (the CLASP modular system) and designed by a capable architect.

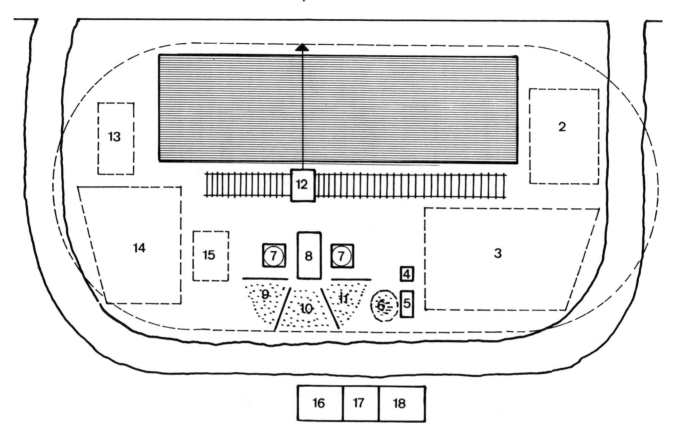

Rationalisiertes Bauen. Die richtige Gestaltung einer Baustelle hat oft eine Steigerung der Produktivität zur Folge.

1 Hauptstraße
2 Siporex-Platten
3 Stahl
4 Mischmaschine
5 Kalk
6 Sand
7 Zementsilos
8 Automatische Mischmaschine
9 Kies 0 bis 7 mm
10 Kies 7 bis 15 mm
11 Kies 15 bis 30 mm
12 1,5-Tonnen-Turmdrehkran
13 Backsteine
14 Bauholz
15 Arbeitsstelle der Zimmerleute
16 Büro
17 Lagerung
18 Arbeiter

Rationalized building. The correct layout of a building site can well cause an important increase in productivity.

1 Main street
2 Siporex slabs
3 Steel
4 Mixer
5 Lime
6 Sand
7 Cement silos
8 Automatic mixer
9 Gravel 0 to 7 mm
10 Gravel 7 to 15 mm
11 Gravel 15 to 30 mm
12 1.5 tons tower-crane
13 Bricks
14 Timber
15 Carpenters working-area
16 Office
17 Storage
18 Workers

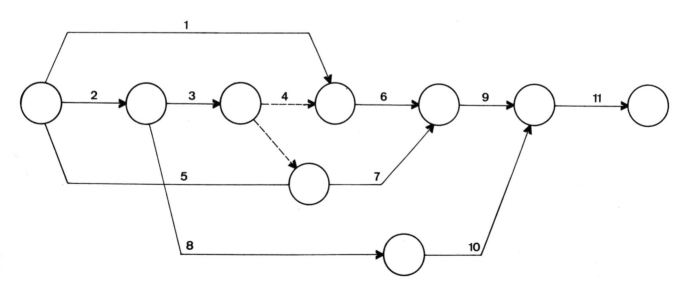

Rationalisiertes Bauen. Die Anwendung guter Planungswerkzeuge steigert die Produktivität, ohne die Technologie zu ändern.

1 Schloß wird geliefert
2 Pläne werden geliefert
3 Tür wird angefertigt
4 Rahmen wird fertiggestellt
5 Scharniere werden geliefert
6 Schloß wird eingebaut
7 Scharniere werden angebracht
8 Türglocke bestellen
9 Tür wird eingehängt
10 Türglocke wird geliefert
11 Türglocke wird installiert

Rationalized building. The utilization of good planning-tools increases productivity without altering the technology.

1 Deliver lock
2 Deliver blueprints
3 Make door
4 Complete frame
5 Deliver hinges
6 Install lock
7 Install hinges
8 Order doorbell
9 Hang door
10 Deliver doorbell
11 Install doorbell

definierte Verbindungstechnik bestimmt werden kann, so wird eine verfeinerte und komplexere Art der modularen Bausysteme geschaffen, die man «Componenting» nennt. Während ein modulares Bausystem nur bestimmte in einem Katalog verzeichnete Teile vorsieht, die der Architekt verwenden kann, werden beim «Componenting» die Entwurfsmöglichkeiten durch die größere Auswahl der zur Verfügung stehenden Bauelemente erweitert. Im Grunde ist es ein gigantisches «Meccano»-System mit Teilen, die bei mehreren, voneinander unabhängigen Unternehmungen erhältlich sind.

«Rationalisiertes Bauen» beruht weder auf einer bestimmten Herstellungsart (Vorfertigung) noch auf einer Maßordnung (modulares Bausystem), sondern auf dem Bemühen, Produktivität und Leistung zu steigern, indem alle erdenklichen Maßnahmen für eine straffe Produktion getroffen und beste Ausnützung von Baustoffen, Ausrüstungen und Arbeitskräften an Ort, also auf der Baustelle und im Produktionsablauf, angestrebt werden. Viele Autoren sehen rationalisiertes Bauen als die fortschrittlichste Art des konventionellen Bauens an, sehen darin jedoch nicht zugleich eine Form der Bauindustrialisierung selbst. In dem von uns definierten Sinn indessen (siehe 1.2) muß rationalisiertes Bauen als eine echte Form der Industrialisierung betrachtet werden.

Zweifellos werden die besten Resultate erzielt, wenn die Rationalisierung bereits am Reißbrett des Architekten beginnt und das Projekt unter Berücksichtigung des *gesamten* Produktionsprozesses, seiner Anforderungen und Einschränkungen geplant wird. Meine praktische Erfahrung hat gezeigt, daß ein Projekt, in dem beträchtliche Mühe und Arbeit auf Planung, Arbeitsvorbereitung, Qualitätskontrolle und Informationsaustausch verwandt wird, sehr wohl in bezug auf Herstellungsdauer, Kosten und Qualität mit Vorfertigung und modularen Bausystemen konkurrieren kann.

Die Ziele der Industrialisierung — Steigerung der Produktion und Leistung — können durch jene Art der Industrialisierung erreicht werden, die ich mangels eines besseren Ausdrucks «ausrüstungsintensive Baustellenproduktion» genannt habe. Bei dieser Art der Industrialisierung wird das Ziel einer gesteigerten Produktivität durch die Anwendung «hochgezüchteter» Werkzeuge erreicht, mit denen man auf der Baustelle leicht vollständige Gebäude errichten kann. In der Praxis hat diese Methode der Industrialisierung bisher wenig Anwendung gefunden, was uns indessen nicht berechtigt, sie außer acht zu lassen. Es ist wegen der geringeren Kosten vorteilhafter, Werkstoffe (Kunstharze, Zement usw.) statt Fertigprodukte zu transportieren. Natürlich braucht nicht alles an Ort gefertigt zu werden; es gibt bestimmte Bauteile, wie zum Beispiel sanitäre Einrichtungen, Fenster usw., die logischerweise in der Fabrik selbst hergestellt und der an Ort gebauten Schale eingefügt werden (siehe Bildbeispiele). Eine erfolgreiche Anwendung dieser Methode der Industrialisierung ist das örtliche Spritzen des Subsystems Innenwand. Diese Methode hat den großen Vorteil, daß sie alle Schallumlenkung und Undichtigkeiten der vorgefertigten Innenwände eliminiert und vielleicht die einzig erfolgreiche Lösung des Problems der festen industrialisierten Innenwände darstellt.

Rationalized building is based not on a production form (prefabrication) or on a dimensional discipline (modular system) but on the attempt to increase productivity and performance by the application of all possible measures for streamlining production, for ensuring the best utilization of materials, equipment and labour on the building site and in the production process. Many authors consider rationalized building as the most advanced form of conventional building but not as a form of building industrialization. However, in view of what I consider to be the essence of industrialization (see 1.2), I am forced to include rationalized building as a true form of industrialization.

Clearly the best results will be achieved when rationalization starts at the architect's desk and the project is designed while keeping in mind the full production process, its requirements and its limitations. Finally my practical experience has indicated that a project where considerable effort has been given to planning, scheduling, quality-control and information-flow, can compete perfectly well in production-time, cost and quality, with prefabrication and modular systems.

The objectives of industrialization (to increase productivity and upgrade performance) can be obtained through the form of industrialization which for lack of a better term I have described as *equipment-oriented site-production*. In this form of industrialization the objective of increasing productivity is achieved by the utilization on the site, of highly sophisticated equipment which can, with little human intervention, produce complete buildings. In practice this form of industrialization has found until now limited application, but this is no justification for disregarding it. In fact, at a first glance, the idea of transporting materials (resins, cement, etc.) instead of finished products is appealing for the obvious reasons of cost. Also the inevitable problem of jointing is avoided to a considerable extent. Clearly, not everything needs to be produced on-site by one or two pieces of equipment; certain parts will logically be factory-produced (sanitary units, operable windows, etc.) and added to the site-produced shell (see pictures). A successful application of this method of industrialization can be found in the spraying on site of the partition sub-system. This method offers the great advantage of eliminating all sound out-flanking and leakages of the prefabricated partitions and possibly offers the only satisfactory solution to the problem of fixed industrialized partitions.

16

Eliot Noyes and Associates, USA.
1　Die Gummiform wird gelegt.
2　Die aufgeblasene Form wird armiert.
3　Das Spritzen von Beton.
4　Die fertige Schale.
5　Der fertige Bau.

Eliot Noyes and Associates, U.S.A.
1　The rubber mould is layed.
2　The inflated mould is covered with steel-reinforcement.
3　Concrete is sprayed.
4　The finished shell.
5　The finished building.

Forschungslaboratorium für Architektur, Universität von Michigan, USA (Forschung von Prof. S.C.A. Paraskevopoulos. Werkzeuge entwickelt von der Dow Chemical Company).
1   Eine Schale wird von einer rotierenden Maschine erzeugt, die Polystyrolblocks legt und zusammenschweißt.
2   Die fertige Schale.
3   Ausschneiden der Fensteröffnung in der fertigen Schale.
4   Der fertige Bau.

Architecture Research Laboratory, University of Michigan, U.S.A. (Research by Prof. S.C.A. Paraskevopoulos. Equipment developed by Dow Chemical Company).
1   A shell is produced by a rotating equipment, which lays and welds polystyrene blocks.
2   The finished shell.
3   Window-cutting in the finished shell.
4   The finished building.

3

4

1.2 Was unterscheidet industrialisiertes Bauen vom konventionellen Bauen?

Als Titel für diesen Abschnitt des Kapitels habe ich die häufigste Frage gewählt, die von den Studenten nach Tagen mühsamer Unterrichtsstunden schließlich gestellt wurde.
Wenn wir wirklich die Definition, die in 1.1 vorgeschlagen wird, annähmen, so müßten wir zu dem Schluß kommen, daß Bauindustrialisierung dann stattfindet, wenn gewisse Produktionsmethoden oder gewisse Formen der Produktionskontrolle (rationalisiertes Bauen) angewendet werden. Und das wäre ein Fehler. Meiner Meinung nach – die heute von einer Anzahl von Unternehmern geteilt wird – ist Industrialisierung hauptsächlich eine Frage der Organisation, außerhalb und innerhalb des auf ihr basierten Unternehmens, während die technische Seite der Industrialisierung nur ein Werkzeug ist, das zu ihrem Erfolg beiträgt. Man könnte sogar behaupten, daß ein industrialisiertes Unternehmen selbst bei Anwendung konventioneller Methoden – wie zum Beispiel dem Aufschichten von Backsteinen – entstehen kann, solange die grundsätzlichen Regeln der Industrialisierung befolgt werden.
Worin besteht nun das Wesen der Industrialisierung? Erste Voraussetzung ist eine subjektiv veränderte Haltung des Bauunternehmers seinem Markt sowohl wie seinen eigenen organisatorischen Möglichkeiten gegenüber. Wenn wir den Gesamtprozeß des Baugewerbes analysieren (siehe auch Kapitel 4), können wir dabei folgende Stufen erkennen:
a) Forschung (neue Methoden, Organisationsformen, Baustoffe, Prozesse, Modelle usw.)
b) Entwicklung (neue Komponenten, Herstellungsvorgänge, Fugenverbindungen, Montagetechniken usw.)
c) Bauentwurf
d) Herstellung (in einer Fabrik oder auf der Baustelle)
e) Planung und Arbeitsvorbereitung
f) Montage auf der Baustelle
g) Planung und Arbeitsvorbereitung der Montage
h) Marketing (mit allen notwendigen Werkzeugen der «Promotion», des Verkaufs usw.)
i) Finanzierung (um dem Käufer die Erwerbung des Produktes zu erleichtern)
k) Landerwerb und -erschließung
l) Behördliche Dienste (Baubewilligungen, Infrastrukturen usw.)
m) Wartungsdienste

1.2 What makes industrialized building different from conventional building?

I have chosen as a title for this sub-chapter the question which students, after a few hours or days of tedious lectures, eventually ask.
Should we in fact make use of the definitions given in 1.1, we would have to accept that the industrialization of building occurs when certain production methods or certain forms of production control (rationalized building) are applied. And we would be wrong.
In my view – shared by now by several industrialists – industrialization is mainly a matter of organization, outside and within the company dedicated to it, while the technical form of industrialization is merely a tool in achieving success. Stretching a point we could state that even a conventional technique (like the one of laying a brick on top of another) could give birth to an industrialized company if certain basic tenets of industrialization were observed.
Of what then does the essence of industrialization consist? Basically, in a change of attitude by the building contractor towards his market and towards his internal organization. When we analyse the full process of the building trade (see also Chapter 4) we identify the following steps:
a) Research (which gives us new methods, organization forms, materials, processes, models, etc.)
b) Development (which gives us new components, production-lines, joints, assembly-techniques, etc.)
c) Building design
d) Production (in a factory or on site)
e) Planning and scheduling of the production
f) Assembly on site
g) Planning and scheduling of the assembly
h) Marketing (with all the necessary tools of advertising, promotion, sales, etc.)
i) Financing (to make it easy for the buyer to acquire the product)
k) Land acquisition and improvement of the land
l) Bureaucratic service (to obtain building permits, infrastructures, etc.)
m) After-sale service
All the above steps have always been and always are respected in conventional or in industrialized building. However, in conventional building the responsibilities are widely dispersed and (as any buyer of a simple product like a family house well knows) nobody is in

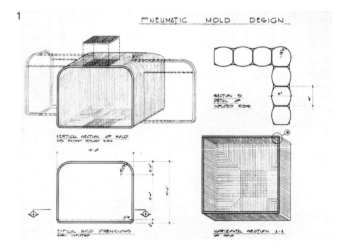

Ein Wohnungsbau aus Schaumkunststoff, entworfen von Studenten der Architektur an der Universität von Kalifornien, USA.
1 Schema der pneumatischen Form.
2 Der Produktionsprozeß auf der Baustelle. (Der Entwurf basiert auf der Entdeckung einer neuen Art von Schaumkunststoff, zu jener Zeit noch nicht im Handel erhältlich.)
3 Mögliche Ausführung eines fertigen Gebäudes.

A plastic-foam housing project, designed by graduate students at the School of Architecture, University of California, U.S.A.
1 Scheme of the pneumatic mould.
2 The site-production process. (The design was based on the discovery of a new kind of plastic-foam at that time not available on the market.)
3 A possible finished building.

Alle genannten Stufen sind beim konventionellen wie beim industrialisierten Bauen immer beachtet worden und werden es auch heute noch. Jedoch sind die Verantwortlichkeiten beim konventionellen Bauen geteilt, was jeder, der ein so unkompliziertes Produkt wie ein Einfamilienhaus kauft, nur zu gut weiß. Niemand ist für den ganzen Prozeß verantwortlich, während beim industrialisierten Bauen eine deutlich identifizierbare Autorität direkt oder indirekt die volle Verantwortung trägt. Außer der integralen Verantwortlichkeit übernimmt der industrielle Unternehmer eine «aktive» Rolle dem Markt gegenüber, während der Unternehmer der konventionellen Methode eine «passive» Haltung übt.

Der konventionelle Unternehmer kümmert sich fast gar nicht um Gesichtspunkte des Marketing oder um Entwicklungs- und Forschungstätigkeiten. Er reagiert nur, wenn ein Bauherr an seiner Tür erscheint und ihn um die Mitarbeit seiner Belegschaft und um die Ausrüstung für die Herstellung eines oder mehrerer bestimmter Gebäude ersucht. Der industrielle Bauunternehmer andererseits ergreift die Verkaufsoffensive, wenn möglich nach einer Marktanalyse; er zeigt sein Produkt werbekräftig an und setzt es schließlich an einen interessierten Käufer ab. Wenn man diesen Vorgang einmal begriffen hat, sieht man den Unterschied zwischen «industrialisiertem» und «konventionellem Bauen» sehr deutlich. Ein Beispiel kann hier helfen. Wenn ich zu einem Tischler gehe und ihn bitte, einen Schreibtisch für mich anzufertigen, dann mag er sehr wohl die allermodernsten Tischlerwerkzeuge benutzen, und doch arbeitet er nach den gleichen Regeln wie sein Vorgänger, der einen Tisch für einen Auftraggeber im dreizehnten Jahrhundert verfertigte. Wenn ich aber in ein Möbelgeschäft gehe und einen Eames-Stuhl kaufe, dann erwerbe ich ein industrielles Produkt, selbst wenn es aus Leder, Holz oder anderen herkömmlichen Werkstoffen gemacht ist. Die völlig veränderte Einstellung gegenüber der Bautätigkeit, die der erfolgreiche Organisator eines industrialisierten Unternehmens unbedingt haben muß, ist oft und mit unbefriedigendem Ergebnis außer acht gelassen worden.

Viele Unternehmer, hypnotisiert von den Anpreisungen der Autoindustrie oder von den vielversprechenden Eigenschaften neuer Werkstoffe, gründen prächtige Fabriken, entwickeln kunstvolle Bausysteme und gehen mit konsequenter Regelmäßigkeit bankrott.

Eine «aktive» Einstellung des Unternehmers ist die charge of the full process, while in industrialized building we find a clearly-identified authority who directly or indirectly runs the whole show. Besides the problem of responsibility the industrial operator takes an "active" attitude towards the market, while the conventional building operator (the contractor) takes a "passive" attitude.

A conventional operator practically ignores the marketing aspects, or the development and research activity: he only reacts when a client appears at his doorstep and asks him for the intervention of his team and equipment for the production of a specific building. The industrial operator, possibly after a market analysis, attacks the market, promotes his product, and eventually sells it to an interested buyer. If we understand this point we clearly see the difference between "industrialized building" and "conventional building". An example may help. If I go to a joiner and ask him to produce a desk for me, the joiner may well make use of the most sophisticated wood-working equipment, yet he still operates according to the same rules of his predecessor, who was producing a table for a gentleman of the thirteenth century. However, if I go to the furniture shop and I buy an Eames chair I am buying an industrial product even if leather, wood and other classic materials are used.

The different attitude required towards industrialized building, by the organiser of an industrialized venture, has been disregarded too often and has always brought unsatisfactory results.

Mesmerized by the production claims of the auto-industry or by the exoteric characteristics of new materials operators tend to create superb factories, develop ingenious construction-systems and with splendid regularity go bankrupt.

An "active" attitude by the operator is a condition necessary for the smooth operation of the production line, and of the assembly, design, and development activities related to it. Without a foreseeable movement of goods, the production will run wild, stocks will grow to an unbearable level, or delays will kill any time-saving advantage that the specific form of industrialization may have made possible.

Curiously enough operators in the field of industrialized building tend also very often to forget the aspect of internal organization. Somehow the idea that a sophisticated process needs to be managed and controlled at all levels by competent and qualified people, is overlooked. The result is that we discover that

2

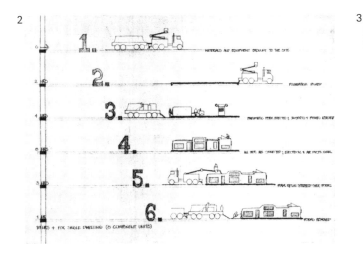

3

notwendige Vorbedingung für das reibungslose Funktionieren des Herstellungsvorganges und für die Montage-, Gestaltungs- und Entwicklungsarbeiten, die damit verbunden sind. Wenn die absatzsichernde Nachfrage nicht vorauszusehen ist, gerät die Produktion außer Kontrolle, das Lager stapelt sich gigantisch auf, oder aber die Vorteile der Zeitersparnis, welche eine spezifische Form der Industrialisierung ermöglicht haben mag, werden durch Verzögerungen wieder wettgemacht.

Manche Unternehmer auf dem Gebiete der Bauindustrialisierung neigen merkwürdigerweise dazu, die innere Organisation zu vernachlässigen. Die Idee, daß ein hochentwickelter Prozeß auf allen Ebenen von kompetenten und qualifizierten Leuten geleitet und kontrolliert werden muß, wird übersehen. Zu spät stellt man dann fest, daß technisch ausgezeichnete Entwicklungen bei günstigen Absatzbedingungen in ernsthafte Schwierigkeiten geraten, weil keine qualifizierten Betriebsleiter angestellt oder keine kompetenten Techniker mit der Produktions- und Montageleitung beauftragt worden waren.

Die Bauindustrialisierung steht und fällt mit der Organisation.

technically excellent solutions operating under favourable market conditions fall into serious trouble because qualified managers have not been employed or qualified technicians were not put in charge of production and assembly.

Organization is the essence of the industrialization of building.

2. Assessment of the Present Situation

2.1 Historical development

Possibly due to well-radicated interests, many parties still choose to look upon the industrialization of building as upon a phenomenon lacking historical background. In reality attempts to introduce industrial production methods into the building trade go well back to the first part of the nineteenth century and considerable successes were registered in the second part of the same century. I feel that the birth-date of industrialization of building can be taken as 1851 with the construction of the Crystal Palace[1].

Possibly from the beginning of the nineteenth century up to the twenties all efforts were devoted to technological innovations. The possibilities of cast-iron and of reinforced concrete, and the introduction of the power-machine, were utilized with very satisfactory results. Besides the already-mentioned Crystal Palace, the Tour Eiffel, the British railway-bridges, the balloon-frame house in the U.S.A., the Halles in Paris are good examples of this period.

From the thirties to the fifties (and in some cases even today) the main efforts of thinkers, architects and some misguided industrialists were devoted to the modular-discipline. These people thought that with a well-developed dimensional coordination most problems of industrialization would be solved, wastages avoided and productivity increased. Sometime towards the middle of this period the "joint craze" developed. The ambition of every student of industrialization was to discover "the joint", possibly "the universal joint". This period probably came to an end with the introduction of the polimerising mastic, in the late sixties, which, offering a reliable joining technique, minimised the problem.

In parallel with the two above-mentioned "fads" we witnessed the strenuous fight between partisans of the "modular systems" — be they "light" or "heavy" — and the partisans of "model" industrialization (defined in 1.1 as prefabrication).

This battle which had the peculiar aspect of being thoroughly incomprehensible has today opened the field for a new "fad": the "componenting". During the last thirty years, when we move from the prehistory of industrialization into the present era, the

---

[1] I think it is useful to refer the reader to a few documents which well cover this particular facet of our subject. *The Prefabrication of Houses* by Burnham Kelly and *L'industrialisation de la construction* by George Van Bogaert (difficult to be found but definitely worthwhile).

Der Crystal Palace von Joseph Paxton in London, 1851. Ein vollständig vorgefertigtes Gebäude aus Gußeisen und Glas.

The Crystal Palace, by Joseph Paxton, London, 1851. A cast-iron and glass, completely prefabricated building.

wenige Unternehmer und Techniker eine einfache Entdeckung: daß die Bauindustrialisierung nur eine Sache der Produktion und des finanziellen Erfolges ist; und allen modischen Theorien zum Trotz gelang es ihnen, einige erhebliche Erfolge zu erzielen.

Während dieser ganzen Entwicklung stellten die Architekten über dieses Thema die faszinierendsten Behauptungen auf, wobei sie die allgemeine Konfusion noch schlimmer machten.

Heute ist die Industrialisierung des Bauens ein wesentlicher und sich ausweitender Teil des Baugewerbes. Wie immer man über sie denken mag, wir müssen uns damit abfinden.

occasional industrialist and technician came to the simple discovery that the industrialization of building was simply a matter of production and financial success, and not withstanding fashionable theories, they managed to achieve considerable success.

All along the architectural profession managed to take the most fascinating positions on the subject, adding to the general confusion.

Today we can safely assume that the industrialization of building is an integral and expanding part of the building trade, and that whatever our opinion is, we better come to terms with it.

2.2 Erfolg und Mißerfolg

Die Analyse der Erfolge und Fehlschläge führt zu einem schwierigen Problem: wo ist die Schwelle, wo Fehlschlag Erfolg wird, festzusetzen? Leider ist der finanzielle Erfolg der einzig mögliche und zuverlässige Maßstab.

Grundsätzlich kann man sagen, daß ein industrialisiertes Unternehmen dann erfolgreich ist, wenn es eine beträchtliche Zeit lang bestehen kann und während dieser Zeit eine angemessene Anzahl von Gebäuden hervorbringt. Dieser Maßstab läßt den qualitativen Aspekt außer acht und setzt uns dem Vorwurf aus, rein merkantil zu denken. Diese Kritik ist aber meines Erachtens nicht berechtigt. Das Problem der Qualität ist von äußerster Wichtigkeit, aber auf unserem Gebiet ist Qualität ohne Quantität bedeutungslos: im Gegenteil, sie ist ein Widerspruch in sich.

Industrialisierung kann nur in einem Kontext der Quantität existieren.

Die meisten Fehlschläge, die mir bekannt wurden, sind auf mangelndes Verständnis der Absatzbedin-

2.2 Success and failure

The analysis of the successes and failures presents us with a difficult problem: How do we distinguish failure from success? Unfortunately the only available and reliable method is that of economic success.

Basically we can say that an industrialization venture has been successful when it has been capable of keeping alive for a considerable period of time, and has managed to produce a meaningful quantity of buildings.

This parameter neglects the qualitative aspect and leaves us open to the accusation of mercantilism. I do not think this criticism is justified. The problem of quality is extremely relevant, but in our field quality without quantity is meaningless: even more, it is a contradiction in terms. Industrialization can only exist in a context of quantity.

Most failures I know of are associated with a lack of understanding of the market requirements. A few, but very few, were caused by an unbelievably naïve design. Unfortunately some failures occurred when

gungen zurückzuführen. Einige wenige wurden durch einen unglaublich naiven Entwurf verursacht. Leider hat es Fehlschläge gegeben, bei denen der Entwurf ausgezeichnet war, zumindest technisch, aber die Leitung nahm an, vielleicht in blindem Vertrauen auf die Qualität des Produktes, daß die Öffentlichkeit automatisch günstig reagieren würde. Zwei solche Fälle in den USA können als Beispiele dienen. Der erste betrifft «General Panels». Eine ausführliche Beschreibung dieses Falles kann in mehreren Veröffentlichungen gefunden werden, wie zum Beispiel in «The Prefabrication of Houses» oder «L'industrialisation de la construction» (siehe Bibliographie).

Für unsere These ist es wichtig, daß der Entwurf hier von Walter Gropius und Conrad Wachsmann stammte. Man braucht wohl kaum zu erwähnen, daß sowohl der Gesamtplan wie auch die Detaillierung tadellos war; der Plan für den Produktionsablauf in der Werkhalle kann auch heute noch fast als ein Musterbeispiel gelten. Und doch endete dieses Unternehmen, das in den vierziger Jahren in Verbindung mit der Kriegswirtschaft und Unterstützung der Regierung der Vereinigten Staaten in Angriff genommen worden war, als vollkommener Fehlschlag. Direkt nach dem Kriege verlor die amerikanische Regierung ihr Interesse, und die Organisation, die sich nicht auf den freien Markt vorbereitet hatte, war einem schnellen Tode geweiht.

Das zweite Beispiel betrifft das «Allside House» in Ohio. In diesem Falle wurde das Unternehmen von einer Privatgesellschaft ins Leben gerufen, der es gelungen war, hochqualifizierte Konstrukteure und Techniker zu engagieren, ohne auf weltberühmte Architekten zählen zu können. Die Herstellungsmethode – auf schaumstoffgefüllten Aluminium-Sandwichpaneelen basierend – war bis ins letzte ausgearbeitet. Das Fertigungsverfahren war fast völlig automatisiert mit einer zentralen Computerkontrolle. Der Gesamtentwurf und die physikalische Eigenschaft der Gebäude übertraf zweifellos alle Konkurrenz. Das Unterfangen endete neun Monate nach Beginn der Produktion. Die Dachorganisation sah sich gezwungen, den drastischen Schritt zu tun und die neue Fabrik zu schließen, um ihr nicht in den Bankrott zu folgen. Wo lag der Fehler? Man hatte einfach das Marketing des Produktes vergessen. Für den durchschnittlichen Verbraucher ist «moderne» Architektur immer noch eine große Unbekannte. Wenn man diesem Mißtrauen die Tatsache zufügt, daß Versicherungsgesellschaften eine höhere Prämie für «moderne» Gebäude verlangen und damit die Finanzierung erschweren (Hypotheken usw.), so sind die Aussichten für ein neues Produkt nicht ermutigend.

Auch in Europa kann man reichlich Beispiele solcher Fehlschläge finden; die Entwicklung verlief ganz ähnlich: ein gutes Design, aber gleichzeitig ein erschreckender Mangel an Marktkenntnissen. Dabei soll aber nicht übersehen werden, daß manchmal auch die Technik versagt hat.

Es ist in der Tat schwierig, ein breites Absatzgebiet vom Wert des Systems zu überzeugen – und ein breites Absatzgebiet ist nötig, um die Industrialisierung lebensfähig zu machen –, wenn die Fugen undicht sind, wenn die Wände sich unter ihrem Eigengewicht krümmen und wenn die Fenster im Gebäude den Wind durch gähnende Lücken hereinlassen.

Probleme dieser Art waren leider Ende der vierziger

the design was excellent (at least technically) and the promotors, possibly blinded by the quality of the product, assumed that the public would automatically react favourably. At least two examples from the United States come to mind. The first example is that of *General Panels*. A full description of this case can be found in several publications such as *The Prefabrication of Houses* or *L'industrialisation de la construction* (see bibliography).

Relevant for our thesis is the fact that the design was the result of the activity of Walter Gropius and Conrad Wachsmann. It is unnecessary to say that both the general conception and the detailing were above reproach: the design for the production flow on the factory floor can still be considered today as being near to perfect. Still this venture, begun in the fourties in association with the war effort, and with the backing of the U.S. government, ended in a complete failure. Immediately after the war the U.S. government lost interest and the organization, unprepared for the open market, was condemned to a rapid death.

The second case is the one of *Allside House* in Ohio. This time the venture was initiated by a private organization which, while not relying on world-famous architects, managed to assemble extremely qualified designers and technicians. The production method – based on a foam-filled aluminium sandwich-panel, was perfectly developed. The production-line was nearly automatized, with a central computer-control. The overall design and the physical performance of the buildings were certainly superior to those of the competitors. The adventure ended nine months after the beginning of production. The mother company had to take the drastic step of closing the new factory to avoid the danger of following the same into bankruptcy. What had gone wrong? Simply that the marketing of the product had been forgotten. To the average public "modern" architecture is still an unknown quantity. When, to this diffidence, we add the fact that insurance companies when dealing with "modern" buildings will ask for a higher premium making it difficult to obtain mortgages and such like, the prospects for a new product are dim.

In Europe similar examples of failures can be found at will. The story is very much the same. A good design, a lack of marketing knowledge. However it may be worthwhile to remember that at times the technique also failed.

It is indeed difficult to convince a wide market (and a wide market is needed to make industrialization viable) when your joints leak, when your walls buckle even under their own load, and when the windows of your building offer yawning gaps to the wind.

This kind of problem was unfortunately rather typical of the industrialized building of the late fourties and early fifties. The fact that such problems have been solved during the past years has not succeeded in lifting a well-grounded diffidence in the public attitude. You have only to look at the reaction of most housewives when you mention the possibility of buying a "prefabricated building".

und Anfang der fünfziger Jahre ziemlich typisch für industrialisiertes Bauen. Daß solche Probleme in den letzten Jahren gelöst worden sind, hat leider das tiefeingewurzelte Mißtrauen der öffentlichen Meinung nicht verringert. Man braucht nur die Reaktion der meisten Hausfrauen zu beobachten, wenn der Kauf eines «Fertighauses» erwähnt wird.

Zusammenfassend kann man die Gründe für Fehlschläge in der Industrialisierung in zwei Hauptgruppen einteilen: erstens Mangel an geeigneter Organisation, besonders äußerer Organisation, die häufigste Fehlerquelle, die uns auch heute noch plagt; und zweitens unzureichende technische Gestaltung, ein sehr seltener Fehler, der aber heute nicht mehr vorkommen sollte.

Summarizing, we can divide the reasons for failure in industrialization into two major groups: the one due to lack of proper organization (particularly external organization), which are in the majority and are still with us today, and the ones, due to a poor technical design, which have always been very few and in any case have practically disappeared.

2.3 Analyse einiger erfolgreicher Unternehmungen

Als Beispiel erfolgreicher industrialisierter Unternehmungen habe ich einen britischen und zwei amerikanische Fälle gewählt.
Das «Consortium of Local Authorities Special Programme» (Spezialprogramm des Konsortiums von Lokalbehörden), abgekürzt CLASP, besteht seit fast zwanzig Jahren und hat in Großbritannien über tausend Gebäude aufgestellt. Um die Ursachen seines Erfolges zu verstehen, muß man eine deutliche Unterscheidung zwischen der technischen und der organisatorischen Seite machen.
Technisch ist CLASP ein Prototyp des modularen

2.3 An analysis of some successful ventures

As examples of successful industrialized ventures, I would like to present one British and two American cases. The "Consortium of Local Authorities Special Programme", in short CLASP, has now been in operation for nearly twenty years, and has to its credit over a thousand buildings in operation in the United Kingdom. To understand the reason for its success we have to distinguish clearly between the technical and the organizational aspects.
Technically CLASP is a prototype of the "modular building system". A set of components is made available to the architects. These components are designed

1, 2 Tupton Hall Comprehensive School, Derbyshire; Architekt: E. Davies, FRIBA, Bezirksarchitekt, 1967. Beispiel eines heutigen Bauwerkes mit dem CLASP-System.
3 FEAL, Italien. Der russische Pavillon auf der Expo 1967 in Montreal. Die Komponenten des Gebäudes wurden in Italien gefertigt, nach Kanada befördert und dort montiert.

1, 2 Tupton Hall Comprehensive School, Derbyshire. Architect: E. Davies, FRIBA, County Architect. An example of a recent building with the CLASP system.
3 FEAL, Italy. The Russian pavilion at the Montreal Expo 1967. The components for the building were produced in Italy, transported to Canada and then assembled.

Bausystems. Ein Satz von Komponenten wird den Architekten zur Verfügung gestellt. Diese Komponenten werden von einer Zentralorganisation, der «Entwicklungsgruppe», entworfen und von ausgewählten Fabrikanten gefertigt. Die Entwicklungsgruppe hat immer eine sehr konservative Einstellung zur Technik gehabt und sorgfältig jedes Experiment mit unausprobierten Baustoffen oder Methoden vermieden. Eine große Ausnahme war die gelenkige, selbststabilisierende Struktur, die von F. W. L. Heathcote, einem der fähigsten Bauingenieure auf unserem Gebiet, eingeführt wurde.

Doch glaube ich mit Sicherheit sagen zu können, daß im Rahmen der CLASP-Organisation jedes andere Design den gleichen Erfolg gehabt hätte. CLASP hat einfach das Problem des Marketings dadurch umgangen, indem die Organisation selbst sich aus einer Anzahl von Lokalbehörden zusammensetzte – zur Zeit über zwanzig –, die das entwickelte System für ihre eigenen Bauprogramme, insbesondere für den Schulbau, verwenden. Auf diese Weise kann CLASP den ausgewählten Fabrikanten leicht einen Voranschlag des Verbrauchs geben. Diese Gruppe von Lokalbehörden hat sich bereiterklärt, die Aufgabe des Entwerfens und der Produktionskontrolle eines Pakets von modularen Komponenten einer zentralen Gruppe zu übertragen, und hat sich gleichzeitig verpflichtet, diese Komponenten zu verwenden. Infolge der Zusammenarbeit mit den Behörden, die in vielen Ländern aus juristischen Gründen nicht möglich ist, wurde 1970 ein Umsatz von zwanzig Millionen Pfund Sterling erreicht. Ein Volumen, das den Einsatz von Forschung, Entwicklung und Ausrüstungsinvestition rechtfertigt. Wenn ein gutes System, kompetente Architekten und Bauunternehmer dazukommen, kann der Erfolg nicht ausbleiben.

CLASP ist auch in anderen Ländern außerhalb Großbritanniens angewandt worden: in Frankreich, Deutschland, der Schweiz, Italien, Israel. Natürlich konnte nur seine technische Seite der fremden Umgebung angepaßt werden, und selbst diese mußte sich den andersartigen rechtlichen und funktionellen Anforderungen fügen. Aber ohne die solide organisatorische Stütze, die in Großbritannien von den Lokalbehörden kommt, war der Werdegang von CLASP im Auslande eine Mischung von Erfolg und Fehlschlag, ähnlich wie in der sonstigen Geschichte der Bauindustrialisierung.

Ein erfolgreiches Beispiel der Industrialisierung in den USA ist «School Construction Systems Development» (Systementwicklung für Schulbau), abgekürzt SCSD. In den sechziger Jahren entwickelte eine Gruppe von begabten Leuten nach dem Muster der britischen Vorbilder – außer CLASP gibt es noch mehrere, ähnlich erfolgreiche Organisationen in Großbritannien – einen organisatorischen und technischen Plan mit Hilfe der «Educational Facilities Laboratories», eines Zweiges der «Ford Foundation». Dieser Plan war dazu bestimmt, einer Anzahl von Distrikten in Kalifornien bessere Schulbauten zu bringen und gleichzeitig den Versuch zu machen, ein wirksames Modell für den industrialisierten Bau zu schaffen.

Mit finanzieller Unterstützung der «Ford Foundation» arbeitete eine Gruppe von Technikern, Architekten, Pädagogen und Administratoren das Konzept einer gutfunktionierenden fortschrittlicheren Schulanlage

by a central organization (the "development group") and are produced by selected manufacturers. Technically the development group has always taken a very conservative attitude and has carefully avoided any kind of experiment with unproven materials or methods. A major exception has been the design of the structure where, due to the activity of one of the most brilliant building-engineers in our field, F.W.L. Heathcote, an articulated, self-stabilising structure has been introduced and utilized.

Still I think we can safely say that any other design would have had the same success within the CLASP organizational set-up. CLASP has simply avoided the problem of marketing the product, the organization itself being formed by several local authorities (at this time over twenty) which utilize the developed system for their own building, in particular school facilities. In this way CLASP can easily assess production quotas for the chosen manufacturers. This group of local authorities have agreed to delegate to a central team the task of designing and supervising the production of a package of modular components. They have also agreed to use these components. Through this bureaucratic artifice (which is not possible in many countries for legal reasons) a market has been created of approximately twenty million pounds (1970), capable of justifying research, development, and production-line investment. Add to this a satisfactory design, qualified architects and contractors, and success becomes a certainty.

CLASP has been utilized in other countries outside the United Kingdom: France, Germany, Switzerland, Italy, Israel. Obviously only the technical content of CLASP was useful in these alien environments, and even the technical content had to be adapted to different legal and functional requirements. But lacking the solid organizational support which is provided in Great Britain by the local authorities, the history of CLASP abroad has been a mixture of success and failure, following the average case-history of industrialized building.

In the United States a successful case of industrialization can be found in the School Construction Systems Development, in short SCSD. Learning from the British experience (beside CLASP several similar organizations have been successfully operating in the United Kingdom) a group of gifted individuals, with the support of the Educational Facilities Laboratories (a branch of the Ford Foundation), developed during the sixties an organizational and technical set-up with the intent of providing better educational facilities to a number of counties in California, and attempted at the same time to establish an operational model for industrialized building ventures.

Having obtained financial support from the Ford Foundation, a group of technicians, architects, pedagogues, and administrators developed the functional content for an upgraded school building. With this document and with official support, SCSD managed

SCSD, School Construction Systems Development, 1966. Äußere und innere Ansicht eines typischen SCSD-Baus. Die Außenwand ist nicht Teil des industrialisierten Systems, sondern wurde der Wahl des örtlichen Architekten überlassen.

SCSD, School Construction Systems Development, 1966. External and internal view of a typical SCSD building. The outside wall was not part of the industrialized system, but was left to the choice of the local architect.

29

aus. Es gelang SCSD mit diesem Dokument und mit offizieller Rückendeckung ein Dutzend Lokalbehörden zu überzeugen, sich zur Errichtung von Schulgebäuden gemäß dem zu entwickelnden System für ungefähr zwanzig Millionen Dollar zu verpflichten. Nachdem der Markt organisiert war, wandelte SCSD das Konzept in einen organischen Satz von Leistungsbeschreibungen um.

Soweit es mir bekannt ist, war SCSD der erste Fall in der Geschichte der Bauindustrialisierung, in dem der ernstliche Versuch gemacht wurde, die Leistung zu steigern. Lange, freie Spannweiten (bis zu 24 m für Decken und 40 m für Dächer), völlig demontierbare und verstellbare Zwischenwände, ein sehr flexibles Subsystem für Temperaturkontrolle und ein differenziertes Beleuchtungssystem waren verlangt. Mit anderen Worten, SCSD ist einer der wenigen Fälle, in denen die Technik des industrialisierten Bauens voll ausgenutzt wird und an denen es deutlich wird, daß überhaupt kein Vergleich mehr mit konventionellem Bauen gemacht werden kann. Man versuche nur einmal, eine Backstein-Zwischenwand zu verstellen!

Auf der Basis der Leistungsbeschreibung wurden mehrere Hersteller von Subsystemen aufgefordert, ein Produkt auszuarbeiten und ein entsprechendes Angebot einzureichen. Von diesen Herstellern wurde je einer in jedem Subsystem ausgewählt (Struktur, Zwischenwände, Beleuchtungsunterdecke, Klimaanlage), und diese Subsysteme wurden den Lokalarchitekten zu genau festgelegten Kosten für den Entwurf der Schulgebäude zur Verfügung gestellt. Zum Schluß wurden auf ähnliche Weise ausgewählte lokale Bau-

to convince a dozen or so local authorities to commit themselves to the building of some twenty million dollars of school buildings with the to-be-developed system. In possession of an organized market, SCSD transformed the functional concept into an organic set of performance specifications.

As far as I know, SCSD has been the first case in the history of industrialized building where a serious attempt has been made to up-grade performance.

Long free spans (up to 75 feet for floors and 125 feet for roofs), thoroughly demountable and relocatable partitions, a very flexible temperature-control sub-system, and a sophisticated artificial-lighting sub-system were required.

In other words, SCSD represents one of the few cases where the technology of industrialized building is properly utilized and it becomes evident that there can be no longer any comparison with conventional building (try to move and relocate a brick partition!).

On the basis of the performance specifications, several manufacturers of sub-systems were invited to develop a product and to present a tender. From among those manufactures one for each major sub-system was selected (structure, partitions, lighting-ceiling, climate control) and these selected sub-systems were made

National Home Ltd., USA, 1970. Zwei Beispiele von Einfamilienhäusern aus Bauelementen der National Home und ergänzt mit lokalen traditionellen Baustoffen.

National Home Ltd., U.S.A., 1970. Two examples of one-family houses produced with components by National Home which are integrated by local traditional materials.

unternehmer mit der Errichtung der Gebäude beauftragt.

SCSD war ein experimentelles Projekt mit einer begrenzten und vorherbestimmten Lebensdauer. Nach Erfüllung der gesetzten Aufgabe wurde die Koordinationsgruppe aufgelöst und das Projekt als abgeschlossen betrachtet. Andere, ähnliche Industrialisierungsprogramme wurden ins Leben gerufen und sind gegenwärtig im Ausführungsstadium (die Toronto-, Montreal- und Floridaprogramme in Kanada und den Vereinigten Staaten und mit erheblich geringerem Erfolg das EDOSS-Programm in Westeuropa).

Wiederum in den Vereinigten Staaten, aber unter ganz anderen Voraussetzungen, gibt es ein weiteres erfolgreiches Beispiel der Industrialisierung, das der National Home. In diesem Falle handelt es sich um ein rein kommerzielles Unternehmen, in dem alle Gesichtspunkte der Gewinnerzielung unterworfen sind. Die technische Seite wird ganz konservativ behandelt, und die architektonische Stilart mag viel zu wünschen übrig lassen. Zweifellos hat die Leitung der National Home die üblichen Fallen vermieden, die so vielen industrialisierten Unternehmen zum Verhängnis wurden.

Ich glaube, daß das wichtigste Charakteristikum dieses Beispiels der Industrialisierung die Vollständigkeit der Leistung ist, welche die Firma anbietet. Einem glatt differenzierten Marktinteresse (Einfamilienhäuser von 15000 bis 75000 Dollar) wird ein Bauentwurf angeboten, der wegen seiner konservativen Züge die beste Aussicht hat, einem größtmöglichen Kundenkreis zuzusagen.

available at a defined cost to local architects for the design of the school buildings. Finally, conventionally-selected local builders were contacted for the construction of the building.

SCSD was an experimental project with a limited and predetermined life-span. At the completion of the pre-established programme the SCSD coordination group was dismembered and the project was given as completed. On similar lines, new industrialization programmes got under way and are at present in their realization phases (the Toronto, Montreal and Florida programmes in North America and with considerably less success the EDOSS programme in Western Europe).

Again in the United States, but operating under completely different assumptions, we find another successful example of industrialization, that of National Home.

The context this time is the one of a purely commercial venture where every aspect is considered from the profit-making angle. The technical content is extremely conservative and the architectural design may well leave a lot to be desired. Certainly the management of National Home avoided the usual traps into which so many industrialized ventures have fallen.

I think the most meaningful characteristic of this example of industrialization lies in the completeness of the service provided by the company. Having identified a wide enough market (one-family-houses starting at 15000 and going up to 75000 dollars), an architectural design has been offered which, due to its conservative aspects, stands the highest chances

Der Prototyp eines Entwurfs – gewöhnlich das Produkt bekannter Architekten – wird von der technischen Abteilung der Firma ausgearbeitet. Produktionspläne werden entwickelt und die wesentlichen Bauteile in der Fabrik gefertigt. Gleichzeitig mit der technischen Entwicklung werden Lokalbauunternehmer und Unternehmungen für die Errichtung und den Verkauf von Fertighäusern herangezogen, mit denen das qualifizierte Personal von National Home die betreffenden Aspekte der Werbung und der Public Relations genau durcharbeitet.

Unter gewissen Bedingungen geht National Home gelegentlich so weit, selbst Land zu kaufen und zu erschließen und das Gebäude vom eigenen Bauunternehmer errichten zu lassen. Außerdem bietet National Home dem zugelassenen Bauunternehmer Finanzierungsmöglichkeiten und schafft somit eines der Haupthindernisse, mit denen neue Unternehmen zu kämpfen haben, aus dem Wege. Wenn der Erfolg der Vorarbeiten fühlbar wird, wird zwischen lokalen Bauunternehmern und National Home ein Arbeitsplan ausgearbeitet, und die Bauteile werden in den eigenen Zubringer-Lastwagen mit Anhängern der Firma zur Baustelle gebracht.

Das Wachstum und die Produktionszuwachsraten der Firma sind zweifellos Beweise für den Erfolg. Vom Standpunkt der Architektur und der Stadtplanung gesehen, ist vieles, was National Home geschaffen hat, nicht wünschenswert; die Methode jedoch hat sich bewährt. Architekten und Planer haben die klare Aufgabe, das Modell des Arbeitsvorganges mit dem menschlichen und sozialen Inhalt zu ergänzen, der dem Produkt die noch fehlenden Werte zufügt.

Während man heute den meisten Architekten zustimmen kann, daß qualitativ Besseres geboten werden sollte, so sehe ich keine Aussicht auf eine Änderung der Situation durch unsere Bemühungen, solange wir uns selbst einfach hinter dem Rauchschirm der Ästhetik und hinter planerischen Ideologien verstecken, ohne uns gleichzeitig die Werkmethode zunutze zu machen, die einige kommerzielle Unternehmen erfolgreich entwickelt haben.

Ich hätte gern dieser kurzen Analyse einiger erfolgreicher Fälle auf unserem Gebiet noch ein Beispiel hinzugefügt, in dem die technische Qualität des Produktes allein den Erfolg gewährleistete. Leider gibt es kein solches Beispiel. Es liegt in der Natur der Industrialisierung, daß der Gesamtprozeß nur dann lebensfähig ist, wenn alle seine Einzelteile richtig funktionieren. Bis wir diese simple Wahrheit gelernt haben, werden noch viele geistige Kräfte der Techniker und Architekten verschwendet und dazu auch noch beträchtliche Geldsummen.

of being accepted by the largest number of potential clients.

The basic architectural design – usually the work of known architects – is then elaborated by the technical department of the company. Production plans are developed and the essential parts of the building produced in the factory. Along with the technical development, local contractors or promoters are acquired to the advantages of constructing and selling prefabricated buildings, and every advertising and public relations aspect of the new project is elaborated with them by qualified National Home staff.

Under certain specific circumstances National Home may go as far as to buy and improve tracts of land and assemble the building through its own fully-owned builder. In addition National Home provides to the licensed builder financing facilities, eliminating in this way a major obstacle common to any new venture. When the effects of the promotional efforts become apparent, a construction schedule is elaborated in conjunction with the local builder and the components of the building are transported to the site by National Home's own fleet of tractors and trailers.

The growth of the company and the production-rates are without doubt a proof of success. From many architectural and town planning points of view the results of the activity of this specific company can well be frowned upon; however, the operational model has proved its validity. It is the clear task of the architects and the planners to add to the operational model a human and social content which would provide the quality lacking in the product.

While I can in fact agree with most architects that better quality should be offered, I cannot see how we stand a chance for a meaningful intervention when we simply barricade ourselves behind the smoke-screen of aesthetical and planning ideology, without taking advantage of the operational vehicle which some commercial operators have successfully developed.

I would like to have added to this brief analysis of some successes in our field, a case where the technical quality of the product has alone been reason enough for success. Unfortunately such a case does not exist. It is in the nature of industrialization that only when all aspects of the process are operating properly does the entire process become viable. Until we learn this simple truism much technical and architectural ingenuity will be wasted, and with it some considerable capitals.

2.4 Die Vorteile des industrialisierten Bauens

Außer der Motivierung des Profits, welche die Haupttriebkraft fast aller Organisationen auf dem Gebiete der Bauindustrialisierung bildet, müssen wir uns fragen, ob die Einführung der Industrialisierung auch eine menschliche Berechtigung hat. Wir können diese Frage unter zwei Aspekten betrachten:
a) der Menschen, die an der Herstellung des Baus beteiligt sind

2.4 The advantages of the industrialized building

Besides the profit-factor, which is the main motivation for most organizations in the field of industrialized building, we have to consider whether the introduction of industrialization has any justification from a human angle. We can approach the question under two perspectives:
a) the people involved in producing the building
b) the people who use the building.

b) der Menschen, die den Bau benutzen und bewohnen.

Wir haben im Baugewerbe während der letzten zwanzig Jahre einen erheblichen Verlust an gelernten Arbeitern festzustellen. Traditionelles Handwerk – zum Beispiel Zimmerleute, Maurer, Spengler –, das die Stütze konventionellen Bauens war, kann die wachsende Nachfrage nicht mehr befriedigen. Gleichzeitig ist der Qualitätsstandard in allen Ländern so stark gefallen – mit Ausnahme der Schweiz, wo «importierte» Arbeitskräfte und die Bereitschaft zu größerem Aufwand einen höheren Qualitätsstandard aufrechterhalten –, daß es heute praktisch unmöglich ist, ein konventionell gebautes Haus in alter solider Qualität zu finden. Da viele Länder eine große Reserve von Arbeitslosen[1] haben, kann das Phänomen wohl nur mit den schweren und unsicheren Arbeitsbedingungen auf der Baustelle erklärt werden. Obwohl diese Arbeit auf der Baustelle immer noch mehr Freiheit und eine unmittelbarere Beteiligung am Gesamtprodukt bietet als andere Arbeitsgebiete, ist für viele die Unbeständigkeit der Arbeitsbedingungen und des Wetters, der ständige Wechsel des Arbeitsortes, Schwankungen in der Nachfrage mit unvermeidlichem temporärem Arbeitsverlust sowie körperliche Gefahren ein Hinderungsgrund, ein qualifizierter Bauarbeiter zu werden. Die Industrialisierung des Bauens bietet den Arbeitern gewisse Vorteile. Ein großer Teil des Produktionsprozesses findet innerhalb einer Fabrik unter besseren Arbeitsbedingungen statt. Das Wachstum einer soliden industrialisierten Bauorganisation mit einem genau festgelegten Produktionsprogramm vermindert die Auswirkungen der Arbeitsschwankungen. Außerdem ermöglicht die Verwendung von Kraftwerkzeugen ungelernten oder angelernten Arbeitern, eine nutzbringende Tätigkeit auszuüben. In vielen Ländern haben Gewerkschaften die Einführung der Bauindustrialisierung bis aufs letzte bekämpft, weil man sich von der Möglichkeit eines Überflusses an Arbeitskräften ernsthaft bedroht fühlte. Zum Glück scheint diese Furcht völlig unbegründet. Vorhandene Statistiken lassen vermuten, daß für die Befriedigung der Nachfrage in den nächsten dreißig Jahren ebenso viele Bauten errichtet werden müssen wie in der ganzen bisherigen Menschheitsgeschichte.

Wir wissen auch, daß wir selbst bei einem Bautempo der Konjunktur der sechziger Jahre lediglich in der Lage sind, die anwachsende Bevölkerung mit Wohnungen zu versehen, ohne die bestehenden Lebensbedingungen zu verbessern. So können wir mit Sicherheit sagen, daß keine Form der Bauindustrialisierung den Arbeitsmarkt des Baugewerbes bedrohen wird, sondern vielmehr die Leistung der Gesamtindustrie hebt.

Ein interessanter und oft übersehener Aspekt der Bauindustrialisierung ist der Verlust der traditionellen Unabhängigkeit und Zähigkeit, die den Bauarbeitern eigen ist und ihnen viel bedeutet. Diese Entwicklung hat schon andere Gebiete menschlicher Tätigkeit betroffen und kann nur durch eine allgemeine Prüfung der Ideologie gemeistert werden. Ganz allgemein scheinen die bisherigen Entwicklungen und die Zukunftsaussichten anzudeuten, daß die Bauindustrialisierung zu

[1] 1970: ungefähr 1 000 000 in Italien, 5 000 000 in den USA, 500 000 in Großbritannien.

In the building trade we have, during the last twenty years, suffered the loss of a considerable amount of qualified workers. Traditional trades like those of carpenters, bricklayers, plumbers, etc., who were the backbone of conventional building, have not been capable of satisfying the growing demand. At the same time, quality standards have been falling in every country (with the possible exception of Switzerland where imported labour and the readiness to pay more have managed to maintain a reasonable standard of quality) to the point that it is now practically impossible to find a satisfactory conventionally-produced building. This phenomenon, since in most countries we still have considerable reserves of unemployed labour[1], may well find an explanation in the hard and unpredictable working conditions of the building-site. While the activity on the building-site still offers more freedom and a more direct involvement in the complete product than is common in other fields, the unpredictable nature of the working conditions is a discouraging factor for many people. The influence of the weather on the activity, the continual changing of the operation theatre, the fluctuation in the demand (with the related periodical loss of employment) make switch to the building activity somewhat unattractive for the normal working man. Also we cannot overlook the physical dangers involved: building is today the most dangerous activity. The industrialization of building has certain advantages for the workers. A great share of the production process is performed in a factory under better environmental conditions. The growth of a serious industrialized building organization, with a well-planned production programme, considerably reduces the consequences of employment-fluctuation. From another angle the introduction of mechanical equipment allows unskilled or semi-skilled labour to be usefully employed. In many countries the introduction of industrialized building has encountered fierce resistance in the trade unions: the possibility of a redundancy of labour was considered a serious threat. Fortunately this preoccupation is thoroughly unfounded. From the available statistical information we know that to satisfy demands we shall have to produce in the next thirty years as many buildings as mankind has produced in all its known history.

We also know that at the building-rate of the booming sixties we will only be capable of providing dwellings to keep up with the natural expansion of the population, without alleviating the existing living conditions. We can, then, safely state that any kind of industrialization will never threaten the employment of the building worker, but simply allow a better performance of the whole industry.

An interesting and often overlooked aspect of the industrialization of building can be the loss, for the workers, of their tradition of independence and toughness to which many are strongly attached. Still this is a development that has already affected other fields of human activity and that only a general reappraisal of our ideology can possibly control. Generally we can say that the past developments and the future outlook indicate that the industrialization of

[1] 1970: about 1 000 000 people in Italy, 5 000 000 in the U.S.A., 500 000 in the United Kingdom.

größerer Berufssicherheit und besseren Arbeitsbedingungen führt.
Wenn wir das Problem von der Seite des Verbrauchers analysieren, so kann man die Erfahrungen der letzten zwanzig Jahre nicht ohne Einschränkungen gutheißen. In einigen Fällen hat die Industrialisierung die Errichtung billiger Wohnungen ermöglicht – zum Beispiel das Einfamilienhaus und das «Mobile Home» in den USA. Aber diese Ersparnis wurde auf Kosten der Qualität erreicht. Niemand kann ernsthaft behaupten, daß das Leben in einer «Mobile Home»-Siedlung ein sozial befriedigendes Erlebnis sei. Ebensowenig haben die pilzartig aufschießenden Siedlungen an der Peripherie europäischer Metropolen (Produkte der konventionellen sowohl wie der industriellen Baumethode) zu einer günstigeren Beurteilung der Industrialisierung beigetragen.
Andererseits hat die erfolgreiche Entwicklung industrialisierter Schulbauten bewiesen, daß bessere Qualität und höhere Leistung bei annehmbarem Kostenaufwand im Bereich des Möglichen liegen. In ganz wenigen Einzelfällen (Krankenhäuser, einige Einfamilienhäuser) hat die Anwendung des Prinzips der Industrialisierung dem Benutzer definitiv geholfen, ein besseres Produkt zu geringeren Kosten zu erwerben. Es gibt offensichtlich keine einfache Lösung dieses Problems. Im Idealfalle kann industrialisiertes Bauen zweifellos bessere Qualität und vielseitigere Gebrauchsleistung zu gleichen oder geringeren Kosten bieten als konventionelles Bauen. Vergleiche mit anderen Industrien (Autos, Fahrräder, Möbel, Kleidung usw.) können diese Behauptung leicht bestätigen. Gebäude aber hängen viel von politischen und weltanschaulichen Umständen ab. Unter falschen Vorbedingungen kann das industrialisierte Bauen ein wirksames Werkzeug zur Weiterführung unmenschlicher Lebensbedingungen werden, und Leute mit Einsicht sollten dies mit allen Mitteln zu verhindern suchen.
Es wäre gut, wenn die Industrialisierung des Bauens die Fehler anderer Industrien vermeiden könnte, in denen die Produktion, anstatt eine Nachfrage zu befriedigen, zum politischen Selbstzweck wird.

building will provide more secure employment, under better physical conditions to the building workers.
If we analyse the problem from the user's viewpoint the experience of the last twenty years cannot be accepted without qualifications. In some cases (as with the one-family house and the "mobile homes" in the U.S.A.) industrialization has allowed the production of inexpensive dwellings. But this economy has been achieved at the cost of serious sacrifices in quality. Nobody can honestly claim that life in a mobile-home camp is a socially satisfactory experience.
On the same lines the mushrooming developments on the outskirts of the European metropolises (produced by conventional as well as by industrialized methods) have done little to justify our belief in the advantages of industrialization. On the other hand the successful developments of industrialized schoolbuildings have proved that better quality and upgraded performances can be achieved while keeping the costs at acceptable levels. In a very few cases (hospitals, some one-family houses) the introduction of the industrialized approach has definitely offered a better value for the user's money. The problem clearly evades a simple answer. Under ideal conditions there can be no doubt that industrialized building can offer a better quality and performance for the same or lower cost than conventional building. Comparisons with other industries (cars, bicycles, furniture, clothing, etc.) definitely support this statement. However, buildings are considerably affected by political and ideological conditions. In the wrong context industrialized building becomes an efficient tool for perpetuating inhuman living conditions and reasonable people should strongly discourage it.
Personally I would like the building industrialization to avoid the traps into which other industries have fallen when production ceases to be a way of satisfying a need to become a mean of self-perpetuation.

2.5 Industrialisierung des Bauens und die Entwicklungsländer

In allen Entwicklungsländern ist die Aufgabe, der Bevölkerung Wohn-, Erziehungs- und Spitalanlagen bereitzustellen, von größter Bedeutung. Es ist naheliegend, daß die Regierungen dieser Länder oft die Einführung der Industrialisierung des Bauens in Erwägung ziehen, um eine schnelle und ökonomische Antwort auf die Bedürfnisse der Bevölkerung zu finden.
Erfahrungen der Vergangenheit haben aber gezeigt, daß industrialisierte Bauten nicht *eo ipso* eine ideale Lösung sind. Industrialisierung ist ein Produkt der westlichen Zivilisation – mit der bemerkenswerten Ausnahme Japans –, und ihre Einführung in eine andersartige Volksgemeinschaft könnte leicht zu einem Bruch der kulturellen Struktur dieser Gesellschaft führen und unerwünschte Folgen haben. Außerdem setzt ein erfolgreiches Industrialisierungsprogramm voraus, daß technische und administrative Kenntnisse

2.5 Industrialization of building and the developing countries

In all developing countries the problem of providing dwellings, educational and health facilities is of primary importance. Logically enough, governments consider at times the possibility of introducing industrialization as a way of finding a fast and economical solution to the needs of the population.
Experience in the past has, however, shown that industrialized buildings are not necessarily the ideal answer to the problem. Industrialization is still a product of the Western civilization (with the notable exception of Japan). Its introduction in an alien society may well cause a rupture of the cultural texture with undesirable results. Besides, a successful industrialization programme presumes that technical and managerial skills are available and that a network of subsidiary manufacturers exists. Obviously an underdeveloped country lacks, practically by definition, these kinds of facilities. Too often then an isolated

vorhanden sind und daß ein Netz von Nebenindustrien besteht. Offensichtlich fehlen in einem Entwicklungsland, wie das Wort schon sagt, diese Voraussetzungen. Nur zu oft gibt es isolierte Unternehmen, die ohne die nötige Unterstützung anderer Tätigkeitsgebiete an Bedeutung für ihre Umwelt verlieren. Man erwäge allein das Problem des Transports: ohne ein angemessenes Straßen- oder Eisenbahnnetz wird der Funktionsradius einer Fabrik für Fertigteile zu bedeutungslosen Proportionen reduziert.

In einigen Fällen wurden Versuche zur Lösung des Wohnproblems gemacht, indem technologisch hochentwickelte Gebäude zu äußerst niedrigen Kosten zur Verfügung gestellt wurden (Kunststoffschalen, mit Kunstharz verpreßte Pappe und plastifizierter Beton usw.).

Ich glaube, daß diese Bauwerke hauptsächlich für den Hersteller und vielleicht für den Lokalunternehmer zufriedenstellend sind. Ihr architektonischer Gehalt ist der einheimischen Kultur fremd, und ihre Beschaffenheit ist gewöhnlich so, daß sie im Ursprungsland nicht aufgestellt werden dürfen. Dazu kommt, daß sie meist ohne Rücksicht auf die Grundregeln der Stadtplanung und ohne Beachtung des Lebensstiles der Bevölkerung errichtet werden. Am Ende wird lediglich der Designer befriedigt sein, der Hersteller macht einen guten Profit, aber die Bauwerke bilden die Saat für ein Elendsviertel.

Eine bemerkenswerte Ausnahme zu diesen unerfreulichen Feststellungen sind die Landschulen, die in Mexiko und Marokko gebaut wurden. In beiden Fällen wurde ein einfaches Design vorgelegt (in Mexiko von einer Gruppe von gut ausgerüsteten Leuten und in Marokko von einem einzelnen französischen Architekten entworfen). Die wesentlichen, leicht zu transportierenden Bauelemente wurden in einer Zentralstelle gefertigt. Regierungsbehörden stellten Strukturteile, vorgefertigte Fenster und Türen und minimale sanitäre Einrichtungen mit Anweisungen für ihren Einbau zur Verfügung. Die Ortsbevölkerung wurde mit der Montage beauftragt und mußte für die Außenwände, die Arbeit am Fundament und die Erdbewegung sorgen. So wurden die Schulgebäude mit limitierten Kosten unter aktiver Mitarbeit der Ortsbevölkerung zur Zufriedenheit aller Beteiligten errichtet.

Ganz allgemein läßt sich das Problem in zwei Hauptkategorien einteilen. Wenn nach Naturkatastrophen eine sofortige Lösung erforderlich ist, müssen uninteressante, aber schnell herzustellende Unterkünfte (Kunststoff-Iglus oder Betonschalen) oft als das kleinste Übel hingenommen werden.

In normalen Zeiten aber bietet eine langsame, die örtlichen Bedingungen respektierende Entwicklung die beste Aussicht auf Erfolg, angefangen mit der Fertigung von primären Baukomponenten wie Fenstern, Strukturteilen, sanitären Einrichtungen usw. Es gibt dann zwar keine aufsehenerregenden Resultate, aber der Prozeß ermöglicht das organische Wachstum einer Nebenindustrie des Bauens ohne Störung einheimischer Gewohnheiten und Techniken. Arbeiter mit nur wenig Erfahrung sind bestimmt in der Lage, die obenerwähnten Bauelemente zu montieren, welche wiederum mit einheimischen Baustoffen verbunden werden können. Örtliche Bauunternehmer können angeregt und unterrichtet, neue und kompliziertere Komponenten können allmählich hinzugefügt werden.

operation, being deprived of the necessary supporting activities, has no meaningful impact on the situation. (Just consider the problem of transport: without an adequate network of roads or railways the operating radius of a factory of prefabricated components is reduced to meaningless proportions.)

In a few cases we have seen attempts to solve dwelling-problems by providing technologically highly-sophisticated buildings at very low cost (plastic domes, plasticised cardboards, plasticised concrete structures, etc.).

I think these buildings are satisfactory mainly to the producers, and possibly also to the local developers. Their architectural connotation is alien to the local culture and their physical quality is usually such that their utilization is forbidden in the producing country. Moreover, they are usually erected without proper consideration to basic town-planning regulations and without the thought for the specific life-behaviour of the population.

The end result is that the ego of the designers is satisfied, the manufacturer is provided with an easy profit and the seeds for future slums are sown. Noticeable exceptions to this rather grim picture are the experiences of Mexico and Morocco in providing rural school buildings. In both cases a simple design (by a well-equipped team in Mexico and practically single-handed by a French architect in Morocco) was adopted, and basic, easily-transportable components were centrally produced. The government authority provided structural members, prefabricated windows and doors, and the basic sanitary equipment, with instructions for their assembly. The local people were in charge of the assembly and had to provide the external enclosure and all foundations and earth-moving work.

With limited costs and with the active participation of the local population school buildings were erected to the satisfaction of all parties involved.

Generally we can break down the problem into two main cases. When natural catastrophes require an immediate solution the utilization of dull but easily-produced dwellings (like plastic igloos or concrete shells) may well have to be accepted as the least evil. In the normal course of events, however, a slow local development, starting with the production of basic, simple components like windows, structural members, sanitary equipment and so on, offers the best chances of success. The results will not be spectacular but the process will permit an organic growth of a subsidiary building-industry without interfering with local habits and techniques. Workers with little training should be capable of assembling the above-mentioned components which should be integrated with local materials. Local builders may be encouraged and educated, and new and more sophisticated components be introduced. From a certain point of view the fact of having to start from scratch can be considered as a blessing. The mistakes of the prefabrication plants of Western Europe and Russia (producers of anonymous rows of dwellings) can be avoided, and a direct connection with the most advanced concepts of componenting be achieved.

Economically the production of essential components makes sense even in countries where labour is overabundant and it offers the possibility for a snow-

In gewisser Weise kann es von Vorteil sein, von Grund auf neu beginnen zu müssen. Die Fehler der Vorfertigungsbetriebe in Westeuropa und Rußland (Hersteller von anonymen Landschaften von Bauten) können vermieden werden, und eine direkte Verbindung mit den fortschrittlichsten Konzepten des «Componenting» ist möglich.

Die Fertigung von unentbehrlichen Komponenten ist wirtschaftlich sinnvoll, selbst in Ländern mit überreichlichem Arbeiterangebot; denn eine solche Produktion löst im Unterschied zu großen Vorfabrikationsbetrieben eine Kettenreaktion aus.

Es scheint logisch zu sein, daß bei der Einführung der Industrialisierung in einem Entwicklungsland das kostbare eigene Investitionskapital nicht an ausländische Lizenzinhaber verlorengehen darf. Eine logischere Methode ist die Nutzbarmachung vorhandener einheimischer Quellen, ergänzt von einfachen und zuverlässigen Hauptwerkzeugen und unter Mithilfe, wenn nötig, einiger ausländischer Experten.

balling effect that big prefabrication-plants will not have.

Finally it seems logical that if industrialization is introduced into a developing country, it should be done without the loss of precious capital to foreign licenseholders. Also it is a more logical way of developing the available local resources, with the possible help of a few foreign experts and the integration of some simple and reliable basic equipment.

3. Architecture and Industrialization

3.1 The situation

In order to understand better the relationship between architecture and industrialized building we should first clarify the meaning of the term "architecture".
I do not mean here the philosophical content of the word, but merely the operational content.
Today the general public assumes that architecture is the result of the activity of an architect, as literature is the result of the activity of writers.
However, the particular feature of architecture is that in most countries nobody can, simply because of his activity, assume the "architect" label. In order to practise architecture, all sorts of regulations and examinations have to be satisfied and only then does an individual receive the license to practise architecture. The situation (apparently designed to defend the public from the activity of poor practitioners, but likely kept alive in order to defend the interests of the "architects") causes some confusion. I think we have to separate the "profession" of the architects from the "function" of the architects. Architecture as a profession allows a group of people to monopolize a field of activity.
As a function, and according to the etymology (*architékton* = chief joiner, master-builder) and historical development of the term, the architect is an individual or a group of persons who are engaged in the creation of buildings. Sometimes uneducated architects achieve highly satisfactory results (the twentieth-century admiration for "spontaneous architecture" and the careful attempt by professionals to recreate the same kind of environment in new towns or in holiday resorts seem to prove that the qualified architect is not the only source of good architecture). At other times less pleasant results, aesthetically, are achieved, as in the "bidonvilles" which originated spontaneously around some of our towns. (All the same it is interesting to note that the most recent sociological studies indicate that in the "bidonvilles" crime rates are inferior to those found in new settlements of professional design.)
However, the difference of architecture as a profession from architecture as an indispensable function of the building activity becomes more relevant when we enter the area of industrialized building.
Official architecture has taken at times a very negative attitude towards industrialization (from the Crystal Palace up to the present day). Some groups of archi-

keit ist von größter Bedeutung, wenn wir in das Gebiet der Bauindustrialisierung eintreten.
Die offizielle Haltung der Architekten gegenüber der Industrialisierung – vom Crystal Palace bis heute – war oft sehr negativ. Andererseits haben einige Gruppen von Architekten die Industrialisierung unterstützt und sich aktiv mit ihr befaßt, vom Bauhaus und der Ulmer Hochschule für Gestaltung bis zu einigen unabhängigen Einzelpersonen.
Diese Beispiele positiver Haltung waren aber nicht nachhaltig genug, um eine Meinungsänderung in der Berufsgruppe hervorzurufen. Man könnte sogar sagen, daß bei weitem die meisten Architekten selbst heute noch die Forderung nach Industrialisierung als einen Eingriff in ihr ureigenes Arbeitsgebiet ansehen. Das Resultat ist – und dieses Phänomen wird zwangsläufig immer schwerwiegendere Bedeutung haben –, daß die Bautätigkeit, mit oder ohne Industrialisierung, auf diese oder jene Weise ohne Berufsarchitekten auskommt. In einigen Ländern, wie zum Beispiel in den USA, werden 95% aller Einfamilienhäuser ohne Berufsarchitekten gebaut. In Ländern des europäischen Kontinents ist es üblicher, den Berufsarchitekten nur zum Zweck der amtlich erforderlichen Unterschrift zuzuziehen. Zwei Fragen drängen sich auf:
a) Warum können Bauunternehmer oder Verbraucher ohne Berufsarchitekten auskommen?
b) Welches sind die Resultate?
Die erste Frage wird ausführlich in 3.2 behandelt. Die Antwort auf die zweite Frage ist leider deprimierend. Die Arbeitsleistung der Bauindustrie ist selbst bei Anwendung konventioneller Techniken so angewachsen, daß sich die Größenordnungen innerhalb von wenigen Jahren verdoppelt haben; Küstendörfer sind zu Großstädten angewachsen. Eine Art «spontane Architektur» hat sich breitgemacht, die nicht im Lebensstil und der Tradition der Ortsbewohner verwurzelt ist und die nicht unter der wachsamen Kontrolle der Gemeinschaft, sondern unter dem Druck gefühlloser Spekulantengruppen entstanden ist.
Wenn Industrialisierung hinzukommt, wächst das Problem, weil es sich dann noch um viel größere Quantitäten handelt. Das Beispiel der «Mobile Homes»

tects have conversely sponsored industrialization and have been actively engaged in it (from the Bauhaus and the Ulm school of design to a few independent individuals).
But these positive attitudes have not been strong enough to create a change of attitude in the profession. In fact, we can say that even today the vast majority of architects looks upon industrialization as an intruder into their vegetable garden. The result has been (and the phenomenon is bound to take ever-growing dimensions) that the building activity, industrialized or not, has, by one method or another, done without the professional architects. In some countries, as in the U.S.A., approximately ninety-five per cent of all the one-family houses built are produced without the intervention of the professional architect. In the continental countries the method uf using the professional architect only for the necessary bureaucratic signature is more popular. At this point we have to consider two aspects:
a) Why have builders or users decided to do without the professional architects?
b) What have been the results?
The first question I would like to answer at length in 3.2. To the second one the answer is unfortunately a sad one. Even under conventional building-techniques the size of the activity reached by the construction industry is such that within a few years our towns have doubled their size; seaside villages have grown into full-scale towns. A kind of "spontaneous architecture" has been born, but its origins are not to be found in the experience and the tradition of the local dwellers (under the vigilant control of the community), but in the activity of unprepared groups of speculators.

1, 2 Neue Quartiere in Rom. Auf Grund der italienischen Gesetzgebung müssen wir annehmen, daß alle diese Bauten das Werk von Berufsarchitekten sind. Photographien: Fernando Cerchio, Rom.
3, 4 Die «Stadtplanung» als Resultat des unkontrollierten Gebrauchs eines industrialisierten Produktes: das «Mobile Home», USA.

1, 2 New quarters of Rome. In view of the Italian legislation we have to assume that these buildings are the work of professional architects. Photos by Fernando Cerchio, Roma.
3, 4 The "town planning" resulting from the uncontrolled use of an industrialized product: the mobile home (U.S.A.).

3

4

(1970: 500000 Einheiten in den USA) und der sie begleitenden Lebensbedingungen spricht für sich.
Die Funktion der Architekten und die Bautätigkeit sind unzertrennlich. Wenn die Mehrzahl der Berufsmitglieder sich aus eigenem Willen von der Bautätigkeit zurückzieht oder wegen Mangels an Vorbildern von ihr ausgeschlossen wird, dann übernehmen andere – Industrielle, Bankiers, Spekulanten, Maurer – die notwendigen Funktionen. Das wäre ganz in Ordnung, wenn diese neuen «Architekten» qualifiziert wären, die Aufgabe zu erfüllen. Leider sind sie es nicht, unsere Städte sind dafür der lebende Beweis.

With industrialization the problem grows because of the quantity factor. The example of the mobile homes (500000 units in 1970 in the U.S.A.) and of the resulting living-environment speaks for itself.
The function of the architect is inherent to the building activity. If the profession in its vast majority chooses to withdraw from the development of the building activity, or because of its lack of preparation is excluded from it, somebody else – the industrialist, the banker, the speculator, the bricklayer – takes over this necessary function. There is nothing wrong with this if these new "architects" were qualified to fulfill the task. Unfortunately they are not, and our towns are living proofs of this situation.

3.2 Die Schwierigkeiten des ausgebildeten Architekten von heute

3.2 The difficulties of the educated architect today

Es gibt mindestens zwei Gründe für das Versagen der Architekten, ihre Pflicht gegenüber der Gesellschaft zu erfüllen und eine annehmbare Umwelt zu schaffen: operationelle Gründe und unzureichende Fachkenntnisse. Der «offizielle» Architekt arbeitet innerhalb der Zunftvorschriften, die hauptsächlich der Verfolgung ihrer eigenen Interessen gewidmet sind, und er ist durch die rechtlichen Begrenzungen seines Berufs gezwungen, als unabhängige Instanz innerhalb des Bauprozesses zu fungieren. Ein Industrieller hat einmal die Situation mit dem klassischen «Dreieck» verglichen: der Bauunternehmer und der Benutzer – für beide ist es ein Risiko, beide stecken Arbeit und Geld in die Sache – und der Architekt, der die Rolle des «Liebhabers» einnimmt – er hat keine Verantwortung, aber den Genuß. Wie anders war der klassische Baumeister und der Renaissance-Architekt, beide Meister des Gesamtvorganges!
Noch größer werden die Schwierigkeiten, wenn der Prozeß der Industrialisierung eingeführt wird. Wenn ein Produkt entworfen werden muß, das mit fortschrittlichen Herstellungsmethoden gefertigt und zu einer festgesetzten Zeit in bestimmten Quantitäten

The reasons for the failure of the architectural profession to fulfill its duty to the society and to accomplish the mission of creating a satisfactory environment are at least of two kinds: operational reasons and lack of knowledge. The "official" architect working within the regulations of static corporative institutions dedicated mainly to the defence of their own interests, is bound by his own legal limitations to operate as an independant authority within the building process. In the word of an industrialist this creates the situation of a classical "triangle": the builder, the user (both of them taking risks and putting effort into the process) and the architect, who acts as the "lover", with no responsibility and all the fun. What a change from the classic "master builder" and the renaissance architect, both real masters of the whole process!
The trouble grows with the introduction of the indus-

Eine neue Stadt für ungefähr 70 000 Einwohner, die auf industrialisierte Weise am Rande von Barcelona errichtet wurde.

A new town for approximately 70 000 people constructed with industrialized methods on the outskirts of Barcelona.

zur Verfügung stehen muß — sei es ein Trennwand-Subsystem oder ein vollständiges Gebäude —, dann wird deutlich, daß die traditionelle Arbeitsweise des «offiziellen» Architekten nicht mehr angemessen ist. Man stelle sich vor, daß ein Flugzeug von einem Fachmann entworfen würde, der die Anforderungen des Benutzers und die Produktionstechniken außer acht läßt und den Entwurf in einem Dutzend Skizzen einer Flugzeugfirma vorlegt!

Es besteht kein Zweifel, daß das Berufsbild des Architekten im Arbeitsprozeß drastisch geändert werden muß, wenn er in Zukunft eine sinnvolle Rolle spielen soll. Die Industrialisierung und die Öffentlichkeit brauchen fähige Architekten. Seine traditionelle Arbeitsweise erschwert dem Architekten das Mitwirken am industrialisierten Bauen, und wenn eine gewisse Größe überschritten wird auch am konventionellen Bauen. Der ausgebildete Architekt von heute hat nicht die Grundkenntnisse, die nötig sind, um einen sinnvollen Platz im Bauprozeß einzunehmen. Die wenigen, die diese Kenntnisse haben, hatten entweder großes Glück mit außergewöhnlich guten Lehrern, oder sie haben die Kenntnisse in der schweren Schule ihrer Erfahrungen gewinnen müssen

Jean Prouvé hat gesagt: «Der Architekt muß ein Hersteller werden.» Das ist richtig, aber mit seiner heutigen Ausbildung sollte der Architekt zu seinem eigenen Schutz lieber jedes Werkzeug vermeiden, das komplizierter als ein Hebelwerkzeug ist.

Manche Leute meinen — und ich schließe mich dieser Meinung an —, daß der Architekt Mitglied einer vielseitig orientierten Arbeitsgruppe werden könnte, in der er die Funktion des Katalysators übernehmen soll. Das ist eine Möglichkeit, aber nur dann, wenn dieser Katalysator fähig ist, mit anderen Mitgliedern der Gruppe eine vernünftige Unterhaltung zu führen.

Indessen ist der moderne Architekt mangels seiner Kenntnisse der wissenschaftlichen Metasprache nicht in der Lage, sich mit anderen Berufszweigen auseinanderzusetzen. Die technischen und technologischen Kenntnisse des durchschnittlichen Architekten sind so niedrig, daß er zu einer potentiellen Gefahr der Gesellschaft wird. Sein Verständnis für die komplexen

trialization process. When a product has to be designed, which shall be produced by sophisticated production methods and made available in large quantities at a specific time (be it a partition sub-system or a complete building), the operational method of the "official" architect becomes ludicrous. Can you imagine a modern aircraft designed by an independent professional, who ignores the user's requirements and production's techniques and defined in a dozen drawings?

There is no question that if the architect is to play a meaningful rôle his operational position must be drastically changed. And let us repeat it: industrialization and the public need qualified architects. But it is not only the operational form which makes the intervention of the architect difficult in industrialized building (or even in conventional building over a certain quantity).

The educated architect of today lacks the basic knowledge to play a meaningful rôle in the building process (the few who possess this knowledge have either been extremely lucky because of exceptional teachers, or have gained their knowledge through their own painful experience).

Jean Prouvé has said: "The architect must become a producer." Certainly, but with his present education the architect had better stay away from any tool more complicated than a lever for his own safety.

Others — myself included — have believed that the architect could well become integrated in a multi-disciplinary team where he would act as a catalyst. This is a possibility, but only if the catalyst is capable of carrying out a meaningful conversation with the other members of the team.

Unfortunately the architect is unable to communicate with other professions due to his ignorance of the scientific language. Technically and technologically the knowledge of the average architect is so low that he cannot be useful to the society.

The understanding of the complex sociological phenomena of a large human conglomerate is at the level of the most naïve political "manifestoes". The understanding of the ways and means that are necessary to

soziologischen Phänomene einer großen Menschenmasse bewegt sich nur auf der Ebene des naivesten politischen «Manifests». Sein Verständnis für die Mittel und Wege, die für die Ausarbeitung eines organischen Programms und Prozesses erforderlich sind, ist einfach nicht vorhanden oder — was noch schlimmer ist — wird von einem mystischen Glauben an die demiurgische Tätigkeit des Computers ersetzt.

Besonders gravierend ist jedoch, daß ihm keine methodologische Grundlage zur Verfügung steht, um sich die erforderlichen Kenntnisse anzueignen, sie miteinander in Beziehung zu bringen und anzuwenden.

Wenn dann noch der ungerechtfertigte Anspruch auf allgemeine Kompetenz dazukommt, dann kann man sich fragen, ob es sich lohnt, den Architektenberuf zu retten oder ob man sich nicht einem neuen zuwenden sollte.

Im Grunde produzieren unsere Architektenschulen im Zeitalter der Biophysik mittelalterliche Alchemisten.

3.3 Die mögliche Stellung des Architekten

Ich habe schon vorher meiner Überzeugung Ausdruck gegeben, daß ein qualifizierter «architektonischer» Einfluß auf den Bauprozeß eine absolute Notwendigkeit ist. Wie kann das erreicht werden? Der erste Schritt könnte sein, die Funktionen im Bauprozeß festzustellen, die vom Architekten erfüllt werden müßten.

Mit Hilfe einiger meiner Studenten habe ich eine Liste dieser Funktionen aufgestellt:
a) Forschung (software design)
b) Entwicklung (hardware design)
c) Projekt (project management)
d) Bauentwurf (building design)
e) Stadt- und Regionalplanung.

Alle diese Funktionen werden bereits von einigen Architekten erfüllt, und meiner eigenen Erfahrung nach besteht eine große Nachfrage nach qualifizierten Spezialisten, vor allem für die ersten drei Funktionen.

Wir müssen für alle diese Funktionen die mit ihnen verbundenen Tätigkeiten feststellen und ebenso die zweckmäßigste Arbeitsmethode.

3.31 Forschung

Architekten haben sich im Laufe der Geschichte mit mindestens zwei Forschungsgebieten befaßt: der Ästhetik und der Technik. Neuerdings hat man jedoch praktisch alle Forschung anderen Berufen überlassen im Glauben, daß die Forschungsresultate durch eine Art Sedimentationsprozeß ihren Weg in unsere tägliche Praxis finden werden. Leider ist es für andere Berufe sehr schwierig, Resultate zu liefern, die in der modernen Architektur gebraucht werden, aus dem einfachen Grunde, weil andere Berufe nicht wissen können, welchen Bedarf die Architektur hat. Ein Beispiel ist die Forschung, die viele neue Kunststoffe hervorgebracht hat, Kunststoffe, die vom Baugewerbe angewandt und wieder aufgegeben werden, weil die Resultate nicht zufriedenstellend sind.

Es gibt aber auch andere Gebiete außer den rein technischen, in denen der Architekt wichtige Arbeit leisten kann. Operationelle Vorgänge müssen analysiert und deren Anwendung in kleinem Maßstab ge-

arrive at the elaboration of an organic programme and process, is simply non-existant or, even worse, is substituted by a mystical belief in the demiurgic activity of the computer.

Finally, and worst of all the architect possesses no methodological basis to obtain, correlate and apply the knowledge which may be required. When added to all this, some architects behave with the tragic attitude of being the only holder of the truth, I wonder if the profession is worth saving or if it may not be more practical to turn to a different profession to fulfill the task.

Basically our schools are producing medieval alchemists in the world of the biophysic.

3.3 The possible position of the architect

I have already stated my belief in the necessity for a qualified "architectural" intervention in the building process. How can this take place? The first step may be to identify the functions in the building process which can be fulfilled by the architect.

I have attempted with some of my students to arrive at this identification and we have established the following list:
a) Research (software design)
b) Development (hardware design)
c) Project management
d) Building(s) design
e) Urban and regional planning.

All the above functions are already fulfilled at least by a few architects, and my own experience has indicated that there exists a large request for more specialists trained in the first three functions.

For each of the above functions we have to consider what would be the content of the activity and under which operational form the activity could best be performed.

3.31 Research

Historically, architects have been involved in research at least on two subjects: aesthetics and technique. Lately, however, practically all kind of research has been left to the care of other professions, in the belief that by some sort of sedimentation the results of such research would find their way into our daily practice. Unfortunately it is very difficult for other professions to deliver the kind of results that are needed for the simple reason that the other professions cannot be aware of the requirements. As an example we can take the research which has produced all kinds of new plastic materials, materials which find their way into the building market only to be rejected because of unsatisfactory performance. The areas of research in which the architect can be extremely useful are more than the merely technical ones. Operational forms need to be analysed and, through application at a limited scale, tested and refined. When we look upon the present forms of tendering, the bureaucratic steps

prüft und verfeinert werden. Wenn man die heutigen Formen der Ausschreibung sieht, die bürokratischen Maßnahmen und Kontrollen, denen ein Projekt unterworfen ist, bevor die Arbeit an der Baustelle beginnen kann, dann merkt man, daß die Dinge sich grundlegend ändern müssen, wenn der dringende Bedarf an mehr und besseren Neubauten befriedigt werden soll.

Methoden der Planung, Arten der Speicherung und Auffindung von Informationen müssen entwickelt werden, um unsere tägliche Arbeit wirksamer und zuverlässiger zu machen. Es müssen Mittel und Wege gefunden werden, selbst die Benutzer in den Planungsprozeß einzubeziehen. Die Wünsche oder vielmehr die Bedürfnisse der Benutzer müssen erforscht und methodisch entwickelt werden, um solche Informationen zu erhalten. Es ist nötig, den Einfluß der Form, Agglomeration, Größe und Lage zu erforschen, um irgendwelche wohlmeinende Fachleute daran zu hindern, neue folgenreiche Mißstände den alten hinzuzufügen. Wir dürfen zum Beispiel nicht vergessen, daß mit heutigen industrialisierten Methoden ein einziger Unternehmer innerhalb eines Jahres eine vollständige Stadt für 50000 Menschen errichten kann.

Die Forschungsgebiete, von denen Architekten für ihre erfolgreiche Arbeit abhängen, um eine wissenschaftlich kontrollierbare Disziplin zu werden, sind zahlreich. In den meisten Fällen müßten auch andere Spezialisten zugezogen werden, zum Beispiel Soziologen, Pädagogen, Technologen, Psychologen usw. Trotzdem aber glaube ich, daß die aktive Teilnahme und oft die Führung durch die Architekten erforderlich ist, um die Resultate in die Praxis umzusetzen. Einige Architekten sind bereits in dieser Art Forschungsarbeit tätig, Männer wie Robertson Ward, USA (Verbraucherbedarf und Werkstoffeigenschaften), Theodore Larsson, USA (technologische Prozesse), Colin Davidson, Kanada (Informationssysteme), Ciro Cicconcelli, Italien (Kosten und Analysenmethode). Die Resultate dieser Studien deuten an, daß unsere Werkmethoden erheblich verbessert werden könnten, wenn sich mehr Architekten dieser Arbeit widmen würden. Auch beweisen diese Erfahrungen, daß Architekten sehr wohl mit anderen Disziplinen zusammenarbeiten und gerade wegen der Natur der Probleme eine führende Rolle übernehmen könnten. Ein anderes Beispiel erfolgreicher Forschung unter der Leitung eines Architekten ist die Arbeit des Laboratoriums für Sozialtechnologie in Japan unter der Leitung von Noriaki Kurokawa.

Es wäre naiv, die praktischen Schwierigkeiten der Forschung auf dem Gebiet der Architektur zu unterschätzen. Kein Einzelarchitekt, keine Firma kann die Kosten einer solchen Forschung tragen. Es gilt, Gönner oder Auftraggeber zu finden und forschungsgerechte Umgebungen und Institutionen zu schaffen. In Anbetracht der steigenden Kritik der Öffentlichkeit an den heutigen Umweltbedingungen sollte die Schaffung solcher Institutionen nicht auf unüberwindbare Hindernisse stoßen.

3.32 Entwicklung

Architekten haben einen viel größeren Anteil an der Entwicklung der «Hardware» gehabt (vollständige Bausysteme und Subsysteme oder Komponenten) als an der Forschung. Auch hier sieht man, daß der Architekt nach Überwindung ursprünglicher Hemmungen,

and controls a project has to undergo before work on site can begin, we can easily see that considerable changes have to be introduced if the outcry for more and better buildings is to be satisfied. Design methods, and forms of collecting and retrieving information have to be developed to make more efficient and reliable our daily work. Also the ways and means of introducing effectively the "users" in the design process need to be worked out. Users' requirements, or better still users' needs, have to be studied, and methods for obtaining this information, developed. Research on the influence of form, agglomeration, size and location, are necessary to prevent some well-intentioned practitioner from adding to the already existing tragedy. (And let us not forget that with industrialized methods a complete town of 50000 people can be produced today by a single manufacturer within a year.) The areas of research on which the architects depends in order to become a scientifically-controllable activity are innumerable. In most cases other specialists will be needed (sociologists, pedagogues, technologists, psychologists, etc.). Still I think the active participation and at times the leadership of the architect is required to insure that the result can be utilized in the normal practice. A few architects have already been engaged in this kind of research (let us mention among others, people like Robertson Ward, U.S.A. [users' requirements and material properties], Theodore Larsson, U.S.A. [technological processes], Colin Davidson, Canada [information systems], Ciro Cicconcelli, Italy [cost and analysis methods]) and their results have indicated that considerable improvements in our daily practice could be achieved if more architects would devote themselves to this field.

Also their experiences have proved that architects can perfectly well cooperate with other disciplines, and assume a leadership rôle by the very nature of the problem. (Another successful example of research led by an architect is the activity of the Social Engineering Laboratory, in Japan, under the leadership of Noriaki Kurokawa.)

It would be naïve to underestimate the practical difficulty of architectural research. No individual architect, or firm, can afford the financial costs of architectural research. Sponsors or clients have to be found, and the correct kinds of research environment and institutions need to be created. I do not think that with the mounting awareness of the public towards the environmental problems the creation of such institutions would be too difficult.

3.32 Development

Architects have been engaged in hardware development (be it complete building systems, sub-systems or components) in much larger scale than in research. A large majority of the available modular building systems have been developed by architects. Here again

Beispiele von «Hardware»-Design. Ein integriertes System von Trennwänden, verglasten Zwischenwänden, Türelementen, eingehängten Möbeleinheiten, Wandschränken und Büroeinheiten. Entworfen für Venesta International Components Ltd., Entwerfer: Anthony Williams and Burles, Component Development Partnership, Photographien: Roger Wood, London.
1   100-mm-verglaste Zwischenwände und feste Trennwand.
2   Eingehängte Möbelstücke an fester Trennwand.
3   Eingefügte Arbeitsplätze an fester Trennwand mit eingehängten Möbelstücken.
4   Ein vertikales Außenwandpaneelsystem aus extrudiertem Spanholz, beidseitig mit farbüberzogenem Stahl verkleidet. Brentwood-ESN-Schule.

Examples of hardware design. An integrated system of partitions, glazed screens, doorsets, suspended furniture, storage-wall furniture and workbays offices. Designed for Venesta International Components Ltd., designed by Anthony Williams and Burles, Component Development Partnership, photos by Roger Wood, London.
1   100-mm-glazed screen and solid area partitioning.
2   Suspended furniture on solid area partitioning.
3   Workbays suspended from solid area partitioning with suspended furniture.
4   A vertical external wall panel system constructed of extruded chipboard faced on two sides with colour-coated steel. Brentwood ESN School.

3

4

die er seiner mangelnden Ausbildung «verdankt», und mit Hilfe der richtigen Informationen andere Fachleute in der Gestaltung der Bau-«Hardware» übertreffen kann. Die Gründe hierfür sind ähnlich wie auf dem Gebiet der Forschung. Ein Berufsingenieur kann vielleicht eine gegebene Struktur mit größerer Leichtigkeit planen, aber nur der Architekt ist in der Lage, die volle Einwirkung des strukturellen Subsystems auf das Gebäude zu überblicken. Architekten überschätzen oft die Komplexität der verwandten technischen Gebiete und verlassen sich allzu gern auf sogenannte Fachspezialisten für den Entwurf von Subsystemen und Komponenten.

Als die Studenten der Universität Washington vor die Aufgabe gestellt wurden, die von technischen Ingenieuren festgelegten Anforderungen zu analysieren, waren sie konsterniert. In enger Zusammenarbeit mit einem qualifizierten Physiker wurde klar, daß mit genügend Zeitaufwand und Mitteln ein viel einfacheres mechanisches Subsystem mit erheblicher Kostenverminderung und Leistungssteigerung entworfen werden konnte.

Es gibt heute einige Architektenfirmen, die sich ausschließlich dem Entwurf von «Hardware» für Bauten widmen. Einige selbständige Architekten haben auch in der Industrie ein befriedigendes Wirkungsfeld gefunden. Beide Arbeitsweisen haben beträchtliche Erfolge zu verzeichnen und haben sich so als annehmbar erwiesen.

3.33 Projektleitung

Wenn ein Industrialisierungsprogramm oder ein Forschungs- oder Entwicklungsprojekt begonnen wird, braucht man zunächst – abgesehen von einem Promoter – einen Projektleiter.

In unserem Tätigkeitsfeld ist es üblich, in jedem erdenklichen Gebiet nach einem Leiter Umschau zu halten, aber nur selten wird die Möglichkeit erwogen, einem Architekten die leitende Funktion zu übertragen. Das rührt daher, daß die Außenwelt vom Architektenstand den Eindruck von Inkompetenz hat. Manchmal ist es der Berufskodex, welcher die Lage verschärft, indem er dem Berufsarchitekten diese Art der Betätigung geradezu verbietet. Neuerdings sind jedoch einige Architekten mit ausgezeichneten Ergebnissen auf dem Gebiet der Projektleitung tätig.

Wenn man sich aber mit heutigen industrialisierten Bauunternehmen beschäftigt, entdeckt man, daß die meisten dieser Organisationen unter der Leitung eines Ingenieurs stehen. Diese Leute haben einfach ihr spezifisches technologisches Wissen beiseite gelegt und «Führungsgewohnheiten» angenommen. Wenn die Bauindustrialisierung die Produktion von Maschinen zum Zweck hätte, wäre nichts dagegen einzuwenden. Da es sich aber darum handelt, eine Umgebung zu schaffen, in der Menschen ein menschenwürdiges Leben führen sollten, wäre mir wohler, wenn ein qualifizierter Architekt die Sache übernähme. Offensichtlich müssen wegen der allgegenwärtigen ökonomischen und politischen Erwägungen Kompromisse gemacht werden, aber ich fühle mich sicherer, wenn diese Kompromisse von jemandem gemacht werden, dem die Interessen des Benutzers am Herzen liegen, statt von jemandem, der seiner Ausbildung nach hauptsächlich an der Herstellung von Werkzeugen interessiert ist.

we find that when the architect overcomes the original inhibitions he has inherited from his own lacking education, and obtains the correct available information, he can out-perform those in other professions in the design of the hardware of building. The reasons are again the same ones given for the research activity. While it is likely that a professional engineer can design with more ease a specific structure, only the architect can see the full implications the structural sub-system has over the whole building design. Also architects tend too often to overestimate the complexity of the related technical fields and too readily rely on so-called specialists for the design of sub-systems and components.

In an exercise carried out at Washington University after analysing the requirements established by service engineers the students were shocked. The collaboration with a qualified physicist indicated clearly that given the time and the means, a far more sophisticated mechanical sub-system could be designed with considerable advantage in cost and performance.

There are today in the world a few architectural firms devoted solely to the development of hardware for building. Also several architects have found it possible and meaningful to operate within the framework of the components industry. Both forms of operation are acceptable and considerable successes have been achieved in both ways.

3.33 Project management

Whenever an industrialization programme or a research or development project have to be started, besides the necessity for a promoter the need arises for a project-manager. It is a common practice in our activity to look for a manager from every possible field, but seldom do the promoters consider the possibility of asking an architect to play the leading rôle.

The fault lies mainly in the unreliable image the architectural profession projects to the outside world. Sometimes professional codes make the situation worse by forbidding this kind of activity to the professional architect. Lately, however, we have seen a few architects entering the field of project management with excellent results.

Actually, when we look into the present industrialization building ventures, we discover that the vast majority of these organizations are managed by engineers. These people have simply put aside their specific technological knowledge and have assumed the "manager" habit. If the objective of the industrialization of building were the production of some kind of equipment, this situation would be above reproach. Being, as is the case, the creation of an environment fit for human life, I would feel happier if a qualified architect were in charge.

Obviously many compromises have to be made for the everpresent economical or political reasons, but somehow I can not avoid feeling safer when these compromises are made by somebody who is interested in the final user rather than by somebody who is, by the very nature of his training, interested in the production tools.

For an architect to play the rôle of a project manager (besides possessing the necessary knowledge) a total commitment to a very different rôle is required. Also

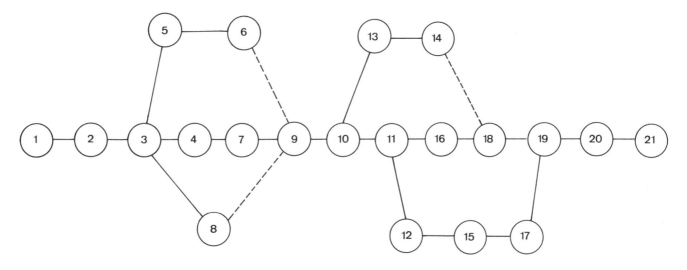

Ein operationelles Schema für das Projekt EDOSS (ein Schulbausystem mit offenem Plan), entwickelt von «Europe Design», einer europäischen Firma von Architekten und Ingenieuren aus verschiedenen Ländern.

	Ereignisse	abgeschlossen
1	Erste Zusammenkunft des Open-Plan-Projektes in Zürich	1. 7.68
2	Entschluß, das Projekt auf seine Durchführbarkeit zu untersuchen	1.11.68
3	Abschluß dieser Untersuchung	2. 2.69
4	Abschluß der Analyse der Hersteller bezüglich Durchführbarkeit	15. 4.69
5	Abschluß der Gespräche zwischen ED-Partnern und Herstellern	15. 4.69
6	Wahl von Herstellern	26. 4.69
7	Plenarsitzung der Hersteller und des ED in Paris	26. 4.69
8	Grundsätzliche Untersuchung der Leistungserfordernisse	26. 4.69
9	Positiver Entschluß der Hersteller	31. 5.69
10	Projektkoordination und Leistungsteam nehmen Tätigkeit auf	15. 6.69
11	Orientierung über Entwicklung der verschiedenen Subsysteme abgeschlossen	15. 9.69
12	Entwicklung der ersten Phase und Systembeschreibung und Gesamtleistung werden festgesetzt	1.11.69
13	PCM (Project Coordination Manager) beendet Auswahl der Marketing-Unternehmer	1.11.69
14	Übereinkunft zwischen Marketing-Unternehmern und Herstellern	30.11.69
15	Abschluß der Entwicklung der Subsysteme	31. 1.70
16	ED-Architekten beendigen Entwurf der Prototypbauten	1. 2.70
17	Fertigstellung der Zeichnungen und Beginn der Produktion der Prototypen	1. 4.70
18	Baubeginn der Prototypen	1. 6.70
19	Fertigstellung der Montage der Prototypen	1. 1.71
20	Abschluß der Korrekturphase	31. 3.71
21	SCC (System Coordination Centre) nimmt Tätigkeit auf	1. 4.71

An operational scheme developed by Europe Design (a European firm formed by architects and engineers of different nationalities) for the EDOSS project (an open plan school building system).

	Events	Date of completion
1	First meeting of the open-plan-project in Zürich	1. 7.68
2	Decision to start the feasability study	1.11.68
3	Completion of the feasability study	28. 2.69
4	Completion of manufacturers' analysis of the feasability report	15. 4.69
5	Completion of bilateral discussions between ED partners and manufacturers	15. 4.69
6	Selection of alternative manufacturers	26. 4.69
7	Plenary meeting of manufacturers and ED in Paris	26. 4.69
8	Basic study of performance requirement	26. 4.69
9	Manufacturers' decision to participate	31. 5.69
10	Project coordination and management team becomes operational	15. 6.69
11	Development brief for each sub-system completed	15. 9.69
12	First stage development and system description and overall performance is established	1.11.69
13	PCM (Project Coordination Manager) has completed the selection of marketing contractors	1.11.69
13	PCM has completed the selection of marketing contractors	1.11.69
14	Agreement between marketing contractors and manufacturers	30.11.69
15	Completion of sub-systems development	31. 1.70
16	ED architects have completed the design of the prototype buildings	1. 2.70
17	Completion of working drawings and start of production for prototype buildings	1. 4.70
18	Start of building operation of prototype buildings	1. 6.70
19	Completion of erection of prototype buildings	1. 1.71
20	Completion of redesign stage	31. 3.71
21	SCC (System Coordination Centre) becomes operational	1. 4.71

Ein Architekt auf dem Posten des Projektleiters braucht außer den nötigen Kenntnissen den vollkommenen Einsatz seiner Kräfte in einer gänzlich neuen Rolle. Außerdem ist in einigen Fällen geistige Beweglichkeit erforderlich. Manchen Architekten mag jedoch die Befriedigung in der Arbeit Entschädigung genug sein für den Mangel an Freiheit und die Komplexität, die diese Art der Betätigung mit sich bringt.

3.34 Bauentwurf und Stadt- und Regionalplanung
Diese Funktionen des Architekten sind so gut bekannt, daß ich hier nicht näher auf sie eingehen will. Nur ein wichtiger Punkt bedarf der Erwähnung: In der Bauindustrialisierung muß die Tätigkeit des «Bauentwurfes» meist innerhalb des Fabrikationsbetriebs oder in engem Zusammenhang mit ihm stattfinden.
Ein Teil der romantischen Architektenfreiheit müßte wohl den Anforderungen der Produktion, der Montage und des Kostenplans geopfert werden. Viele junge Architekten weigern sich immer noch, auf diese Weise dem Produktionsprozeß einverleibt zu werden. Vom Standpunkt ihres egoistischen Interesses ist die Haltung verständlich; für die künftigen Benutzer des Produktes könnte sie einfach unnötige Kosten, Mangel an Qualität und lästige Verzögerungen bedeuten.

in some cases considerable mobility becomes necessary. However, the satisfaction that production gives may well compensate, for some architects, the lack of freedom and the complexity that this form of architectural activity demands.

3.34 Building design and urban and regional planning
These functions of the architect are so well-known that I do not need to dwell upon them. Only it may be relevant to mention that with industrialized building the activity of "building design" may well have to take place in close association with or within the manufacturing organization itself.
A nice slice of the romantic architect's freedom may have to be sacrificed to the requirements of production, assembly and financial schedule. Many young architects still tend to refuse this integration into the production line. From their own egoistic interest the position is easily understandable; for the future user of the product it may simply mean unnecessary cost, lack of quality and annoying delays.
Conversely the functions of urban and regional plan-

Andererseits können die Funktionen der Stadt- und Regionalplanung nur dann richtig durchgeführt werden, wenn der Architekt ein Teil der politischen und Entscheidungs-Maschinerie wird, mit nachdrücklicher Ausnahme der Forschung auf beiden Gebieten, in denen weitestgehende Autonomie wünschenswert und notwendig ist. Schöne Stadtplanungsprojekte, die von unabhängigen Architektenfirmen entwickelt werden, haben wenig Aussicht auf Verwirklichung. Es mag oft lästig scheinen, politischen oder administrativen Organisationen beizutreten, aber es ist genau so unerfreulich, seine Zeit für unerreichbare, zeitraubende Luftschlösser zu verschwenden.

3.35 Zusammenfassung

Die Analysen der bestehenden Lage, die verbessert werden muß, gehen von einer Hauptvoraussetzung aus: Der Bauprozeß braucht die Mitwirkung ausgebildeter Architekten. Hieraus ergeben sich folgende Notwendigkeiten:
a) Der Architekt muß eine möglichst gute Berufsschulung haben
b) Der ausübende Architekt muß mit begrifflichen und physischen Werkzeugen für seine Arbeit ausgestattet werden
c) Der Architekt muß die Rolle des passiven, ad-hoc-engagierten Luxusgegenstandes gegen die wesentliche und wenn möglich führende Teilnahme an einem komplexen Prozeß eintauschen.

Eine politische Kontrolle «des Volkes» über die Tätigkeit des Architekten ist durchaus erwägenswert. Zweifellos müßten verschiedene, den jeweiligen politischen Systemen angepaßten Methoden angewandt werden. Jedoch muß besonders bei den heutigen Produktionsmethoden eine Art der Kontrolle während der Planungs- und Durchführungsphasen eingeführt werden.

ning can only be properly fulfilled when the architect is integrated within the political and administrative decision-making machinery (with the strong exception of research in both fields, where the widest autonomy is desirable and necessary). Town planning projects elaborated by independent firms stand the same chances of realization as a paper butterfly of reaching the moon. It may well be unpleasant to join political or administrative organizations, but it is even more futile to spend one's own time-developing schemes which will never be realized.

3.35 Conclusions

All the above analyses of a situation which exists and which needs to be improved upon, stem from a basic hypothesis: the building process requires the intervention of a prepared architect. From this we derive several needs:
a) To educate as well as possible this professional
b) To provide the working-architect with conceptual and physical tools with which to work
c) To change the position of the architect from a passive, ad-hoc hired luxury-item into an integral and possibly leading part of a complex process.

Politically the control of "the people" over the activity of the architect should be well considered. Clearly, according to each political system, different methods may have to be used. However, particularly with the present production methods, some form of control at the conception and during the realization stage must be introduced.

4. Organizational Concepts

4.1 Who has been behind the industrialization of building?

All kinds of individuals and organizations, with the most varied motivations, have initiated industrialized-building ventures. At times the advantages of prefabrication in time-saving have been at the origin of a particular project, coupled — as in the case of the Crystal Palace — with the availability of a new technology.

Social awareness has caused some well-meaning late nineteenth-century builders to introduce in the United Kingdom "ferro-concrete" prefabricated houses for workers (going bankrupt in the process). The gold-rush and subsequent colonization of the Western United States, combined with the new possibilities provided by the wood-working power-machine and a normal business acumen were the reasons for the introduction of the prefabricated balloon-frame houses. A certain political and ideological conception were at the origin of the Frankfurt industrialization efforts of the twenties.

In more recent times we can identify at least three major forces behind the industrialization ventures:
a) The desire to make money
b) Testing an intellectual scheme
c) A specific need for a certain kind of dwellings.

The first motivation we can ascribe to most of the ventures originated by existing conventional contractors and manufacturers of building-hardware. However, it may be dangerous to over-simplify. Some manufacturers or contractors entered the field of complete industrialized buildings only because they saw in the sale of the completed buildings a good vehicle for the utilization of their own specific product or material. To such cases we can ascribe the products of industries like U.S. Steel, British Steel, Durisol, Hoesch, several window- or curtain-wall manufacturers, and people from the wood-working industry.

Obviously this kind of firm enters the field with a completely different attitude (and economical calculation) to that of organizations which see in the industrialization of building a rewarding activity in its own right.

To this second group I think we can ascribe organizations like National Home, Constructions Modulaires, IGECO, IPI, and hundreds more.

While in both cases the motivation is economical, these organizations operate differently and particularly

Motivierung in beiden Fällen ökonomischer Natur ist, sind die Arbeitsweisen verschieden, und es liegt auf der Hand, daß besonders die technischen Besonderheiten ihrer Produkte äußerst verschiedenartig sind.

Es gibt Spezialfälle, in denen Organisationen das Gebiet aus «Prestige»-Gründen wählen. Diese Fälle kommen häufiger vor, als man in der sogenannten «kalten» Welt der Industrie erwarten sollte. Ich habe Bauherren gehabt, die ich von dem Wagnis eines beträchtlichen industrialisierten Unterfangens abhalten mußte, als ich entdeckte, daß ihre einzige Motivierung der Wunsch war, die Konkurrenz zu beeindrucken.

Der Wunsch, eine intellektuelle Anschauung durchzusetzen – technologischer oder technischer Natur – ist eine typische Motivierung für industrialisierte Unternehmen gewesen. Es muß aber hier gesagt werden, daß die meisten dieser Unterfangen zu einem Debakel führten. Oft sind Architekten, Ingenieure oder «Erfinder» die Urheber solcher Schemas. Es gibt Unternehmer, die ohne die solide Basis des Profitmotivs im Vertrauen auf ein besonderes Paneel, eine Fugenverbindung, eine hochgezüchtete Ausrüstung oder ein neues Modul-«Spiel» vorerst auf den Gewinn verzichten, in der Hoffnung auf einen künftigen Riesenerfolg. Der Riesenerfolg stellt sich nie ein, und viele gute Ideen haben ein trauriges Ende genommen.

Heute wurden Organisationen wie CLASP, SEAC, SCOLA und die Industrialisierungsprogramme der russischen Regierung sowie die mexikanischen und marokkanischen Landschulprogramme und noch einige andere ins Leben gerufen, um einen spezifischen Bedarf zu decken. Da gewöhnlich eine Behörde oder eine halbamtliche Organisation die treibende Kraft ist und somit den Markt kontrolliert, sind diese Unternehmen erfolgreich, nicht nur nach dem Maßstab der Produktivität, sondern auch in qualitativer Hinsicht. In den meisten Fällen war eine sorgfältige Forschung möglich, Neuerungen konnten eingeführt werden, da es keine Bindungen an bestimmte Herstellungstechnologien gab. Außerdem, und das ist wichtig, wurden die Bedürfnisse des Benutzers mit mehr Sorgfalt und ohne Absicht der Manipulation erforscht, die bei rein kommerziellen Unternehmungen üblich sind.

Ein Überblick über die Protagonisten, die sich in den letzten zwanzig Jahren mit der Industrialisierung des Bauens befaßt haben, zeigt, daß sie einem breiten Berufsspektrum angehören: Bankiers, Versicherungsgesellschaften, Hersteller von Rohmaterial, Hersteller von Fertig- oder Halbfertigprodukten, Bauunternehmer, Promoter (von Feriensiedlungen usw.), Architekten, Ingenieure, Designer, Behörden, Militär, einige Gewerkschaften, Spekulanten und sogar einige in Genossenschaften organisierte Benutzer.

the technical characteristics of their products tend for obvious reasons to be extremely different.

A special case is represented by the organizations which enter the field for "prestige" reasons. These cases exist more often than would be expected in the theoretically "cool" world of industry.

I have had clients who had to be refrained from embarking on sizeable industrialization ventures when I discovered that the only reason behind their interest was to "impress" their competitors.

Proving an intellectual scheme (based mainly on some kind of technological or technical improvement) has been a classical motivation for industrialized ventures. It may well be stated here that most of such ventures have ended in major catastrophes.

Architects, engineers, or "inventors" have been the originators of such schemes. Lacking the solid basis provided by profit-motivation and relying on the success of a certain panel, joint, piece of equipment or modular game, the promoters have too often been willing to operate "at cost" in the hopes of a future major breakthrough. The breakthrough regularly failed to materialize and many good ideas have come to a sad end.

The satisfaction of a specific need has been in recent times the motivation of organizations like CLASP, SEAC, SCOLA, the industrialization effort of the Russian government, the Mexican and Moroccan rural school programmes, and a few more.

A public or semi-public authority normally being the promoter, and thus in control of the market, these ventures have usually been successful not only from a mere production-rate measurement, but also from a qualitative point of view.

In most cases careful research has been possible, and being free from conditioning ties to a manufacturing technology, innovations have been possible. Also, and more important, the requirements of the final users have been studied with more care and without the intention of manipulating them, as is usually inherent to most purely commercial ventures. A review of the physical protagonist of the industrialization of building in the past twenty years covers a very wide spectrum of professions: bankers, insurance companies, manufacturers of raw materials, manufacturers of finished or semi-finished products, building contractors, promoters (holiday resorts, etc.), architects, engineers, designers, public authorities, the military, some unions, speculators, and even a few "users", grouped into cooperatives.

4.2 Wie werden industrialisierte Unternehmen betrieben?

Erfolgreiche (und erfolglose) industrialisierte Unternehmen haben viele verschiedene organisatorische Schemas für ihre Betriebsform gewählt. Für die gewählte Betriebsform ist eine A-priori-Entscheidung für ein Design oder ein Produktionsschema wichtiger als das Vorhandensein des Profitmotivs. Es mag nütz-

4.2 How do industrialized ventures operate?

Successful (and unsuccessful) industrialization ventures have chosen to operate under quite a few organizational schemes. Most important for the operational form chosen has been an aprioristic commitment to a design or a production scheme, rather than the presence or lack of profit-motivation. It may be useful

| Forschung / Research | Entwicklung / Development | Landerwerb und -erschließung / Land acquisition and land improvement | Bauentwurf / Building design | Promotion / Promotion | Marktforschung / Marketing | Behördliche Dienste / Bureaucratic service | Finanzierung / Financing | Produktionsplanung / Production scheduling | Produktion / Production | Montageplanung / Assembly planning | Montage / Assembly | Wartungsdienst / After-sale service |

Die Phasen des industrialisierten Bauprozesses / The general stages of the industrialized building process

lich sein, die Analyse mit den beiden Extremen der Organisationsarten zu beginnen:
a) Die Organisation, die «alles macht»
b) Die Organisation, die «nichts produziert».
Tatsächlich kenne ich keine Organisation, die wirklich «alles macht», einfach weil gewisse Bauelemente immer vorteilhafter von anderen Spezialherstellern gekauft werden, zum Beispiel Türen, Eisenwaren, sanitäre Anlagen, Fenster usw. Jedoch beschreibt der Ausdruck weniger das Unternehmen, welches alle Bauteile fertigt, als vielmehr das Unternehmen, welches die eigene Herstellungsmethode entwickelt — Vorfertigung, modulare Bausysteme usw. — und danach die Mehrzahl der Bauelemente herstellt und den fertigen Bau selber oder durch spezialisierte Makler verkauft. Eine solche Firma braucht natürlich ein großes investiertes Kapital, sie muß auf die Produktion eines besonderen Gebäudetyps spezialisiert sein (Schulen, Wohnbauten, Lagerhäuser usw.), und sie ist an ein bestimmtes geographisches Gebiet gebunden, das vom ökonomisch tragbaren Transportradius der Bauelemente abhängt. (Die Firma Nachbarschulte in Deutschland ist ein typisches Beispiel, siehe Kapitel 5.)
Wenn eine geographische Markterweiterung lohnend erscheint, werden gewöhnlich zusätzliche Herstellungsanlagen geschaffen, oder die Lizenz für das «System» wird einem neuen Unternehmer übergeben, mit oder ohne wirtschaftliche Beteiligung der Originalfirma. In Europa gibt es viele Beispiele dieses Vorgehens: Larsen und Nielsen, Camus, Coignet, Balency und FEAL; alle gehören dieser Gruppe an.
Wenige Firmen exportieren wegen der besonderen Eigenschaften ihres Produktes — Leichtgewicht, Transportfähigkeit, Kapitalintensität der Bauelemente — Fertigbauten. Allgemein wird das von den importierenden Ländern nicht gern gesehen, aber außergewöhnliche Umstände, wie zum Beispiel Naturkatastrophen oder besondere lokalpolitische Zwecke, machen das Exportgeschäft zuzeiten recht einträglich.
Häufiger ist der Verkauf einer «Know-how»-Lizenz an einen ausländischen Unternehmer. Diese Lizenz wird gewöhnlich in Verbindung mit besonderer Produktionsausrüstung und technischer und administrativer Beratung verkauft. Das Lizenzgeschäft hat während

to begin our analysis with the two extreme operational forms of the spectrum:
a) The organization which "does it all"
b) The organization which "produces nothing".
Practically, I do not know of any organization which really "does it all" for obvious reasons: certain components of the building are always better bought from other specialized manufacturers (doors, hardware, sanitary equipment, windows, etc.). However, more than the production of all the components of the building, the definition describes the kind of company that develops its own production method (prefabrication, etc.), then goes on to produce the majority of the components and sells the product to the user or passes on this responsibility to a specially-organized agent.
This kind of company is naturally characterized by considerable capital investment, specialization in the production of a particular kind of building (educational, housing, warehouses, etc.) and by a limited geographic area of activity, determined by the economic transportability radius of the components. (Nachbarschulte from Germany is a perfect example of this: see chapter 5.)
Usually when a geographic expansion becomes profitable new production facilities are created or the "system" is licensed to a new operator, with or without economic participation of the original company.
Examples of this form of operation are common in Europe: Larsen and Nielsen, Camus, Coignet, Balency and FEAL, all generally fit into this group.
A few companies, because of the particular characteristics of their product (light-weight, transportability, value of the components) engage in the export-activity of finished buildings. Generally this kind of activity is frowned upon by the governments of the importing countries, but unusual circumstances (like natural catastrophes) or local political expedience allow at times rather profitable export-operations.
More common is the sale of a "know-how" license to a foreign operator, a license usually sold in connection with specific equipment and a general technical and administrative supervision. The "license" business has taken, during the last ten years, considerable dimensions and not uncommon is the case of a company which operates at a loss in its original

Ausgeführt von der industrialisierten Firma	Ausgeführt von der Firma oder dem Bauherrn	Ausgeführt vom ausgewählten Hersteller	Ausgeführt vom ausgewählten Bauunternehmer
Done by the industrialized organization	Done by the organization or the client	Done by the selected manufacturer	Done by the selected contractor
Entwicklung Promotion Marktforschung Auswahl der Hersteller Montageplanung Wartungsdienst Development Promotion Marketing Selection of manufacturers Assembly planning After-sale service	Bauentwurf Behördliche Dienste Finanzierung Landerwerb Building design Bureaucratic service Financing Land acquisition	Produktion Produktionsplanung Production Production scheduling	Montage Montageplanung Assembly Assembly planning

Die allgemeine Organisationsform einer Firma, die «nicht produziert». / The general operational form of the "non-manufacturing organization".

der letzten zehn Jahre beträchtliche Dimensionen angenommen, und es gibt oft Fälle von Firmen, die im eigenen Lande Geld verlieren, aber ihre Bilanz dank den Lizenzeinnahmen retten.
Firmen, die ihre eigenen Komponenten herstellen, bieten gewöhnlich ein stark genormtes Produkt an, weil hier das in die Produktionsanlage investierte Kapital sehr groß ist und deshalb zu einer Einschränkung der Gestaltung führt.
Ganz anders ist die Art des Vorgehens bei Firmen, die «nichts produzieren». Hier ist die Definition wörtlich gemeint. Diese Art der Firma, die meist für rein spekulative Zwecke gegründet wird, entwickelt eine bestimmte Baumethode, die einer vorher ausgearbeiteten Marktanalyse entspricht.
Gleichzeitig mit der technischen Entwicklung werden geeignete Hersteller von Baukomponenten herangezogen, und mit ihnen wird die Detaillierung der Komponenten verfeinert. Kosten, Lieferungstermine, Qualitätsnormen werden festgesetzt. Zur gleichen Zeit setzt die Werbetätigkeit ein, die je nach Land auf Architekten, lokale Bauunternehmer, öffentliche Behörden, den Verbraucherkreis oder auf alle gerichtet wird.
Falls einer dieser potentiellen Auftraggeber das vorgeschlagene «System» annimmt, kann die Firma auf mehrere Weisen vorgehen. Manchmal schließt sie den Vertrag ab und übergibt einem konventionellen Bauunternehmer die Montage der von Fabrikanten hergestellten Komponenten. Oder aber ein geeigneter Bauunternehmer übernimmt die volle Verantwortung für die Errichtung des Gebäudes und kauft unter bestimmten legalen Bedingungen die Baukomponenten direkt vom Fabrikantenring.
In einer anderen, verfeinerten Methode, die aber organisatorisch schwieriger ist, werden die Komponenten direkt vom Hersteller auf die Baustelle geliefert und montiert. Die organisierende Firma überwacht diesen Arbeitsgang und gibt spezifische Baustellenarbeiten wie Aushub, Erdbewegung usw. in Unterverträgen an die betreffenden Unternehmen aus.
Diese Art des Vorgehens wirft verschiedene Fragen auf:
a) Warum reagieren die Auftraggeber auf die Werbung einer Firma, die «nichts produziert»?
b) Kann eine solche Organisation Erfolg haben?
c) Kann das Endprodukt zufriedenstellend sein?

country and closes its book with a profit due to the royalties from foreign licensees.
Technically the companies which produce their own components usually offer a rather standardized product due to the conditioning factor of their considerable capital investment in equipment.
A completely different approach is taken by the company "which produces nothing". In this case the term can be taken literally, from a physical point of view. This kind of company, usually formed for purely speculative reasons, develops a certain construction method in accordance with the previous results of a market-analysis.
In parallel with the technical development, suitable existing component-manufacturers are contacted and selected. Cost, delivery terms, quality-standards are developed. Along with the above activities a promotion campaign is started, directed, according to the country of operation, towards the architects, the local contractors, the public authority, the private users, or all of these people at the same time.
Assuming that one of these potential clients accepts to utilize the developed "system" the company may choose to operate in several forms. At times the company contracts the job and then subcontracts to a local builder the task of assembling the components produced by the "manufacturer ring". Alternatively a suitable contractor takes direct responsibility for the assembly of the building and by means of suitable legal arrangements buys the components from the manufacturer ring.
In a more sophisticated (but organizationally more difficult) operational form, the component manufacturers deliver their own products "assembled" on site. The organizing company supervises their activity and subcontracts to minor contractors the specific site-activity of excavation, earth movements, etc. This kind of operation raises several questions:
a) Why do clients respond to the promotion of a company which "produces nothing"?
b) Can such an organization be successful?
c) Can the product be satisfactory?
It is in fact usually difficult to convince an unprepared client to buy a building from someone who does not have his own production- and assembly-facilities. Still, a well organized company can easily prove to the

Es ist in der Tat im allgemeinen schwierig, einen unvorbereiteten Auftraggeber zu überzeugen, ein Gebäude von jemandem zu kaufen, der weder eine eigene Fertigungsanlage noch eine Montageausrüstung hat. Und doch kann eine gut organisierte Firma einen Auftraggeber leicht von den Vorteilen eines bis aufs letzte durchgearbeiteten Produktes überzeugen, das nicht durch bestimmte Baustoffe und Herstellungsmethoden bedingt ist und dessen Kosten durch Masseneinkauf niedriger gehalten werden. Wenn der Auftraggeber ein Architekt ist, liegen die Vorteile einer technisch durchgearbeiteten Bauweise auf der Hand: die Sorge um Details ist ihm abgenommen. Für den Bauunternehmer sind verläßliche Lieferung und Zeitersparnis überzeugende Argumente.

Tatsachen haben bewiesen, daß ein solches Vorgehen gute Ergebnisse haben kann, besonders in Westeuropa, wo Firmen wie Brockhouse-Systembau in Deutschland, Constructions Modulaires in Frankreich und Servicio Técnico de Construcciones Modulares in Spanien sehr erfolgreich sind. Es ist aber interessant zu beobachten, daß es kein Beispiel ähnlicher Unternehmen für Wohnbauten gibt.

Von einem qualitativen Standpunkt aus bietet dieses Vorgehen hauptsächlich den Vorteil großer Flexibilität. Und eine Organisation kann nach Belieben neue Subsysteme in ihre allgemeine Konzeption aufnehmen. Nur der ein- oder zweijährige Vertrag mit den Herstellern bildet eine Einschränkung.

Natürlich hat ein solches Vorgehen auch gewisse Nachteile. Die ernsthafteste Schwierigkeit könnte durch einen unerwarteten Verkaufserfolg verursacht werden, und somit wäre es für die Hersteller unmöglich, die Subsysteme innerhalb der vereinbarten Frist zu liefern. Da Zeitersparnis eine der Hauptattraktionen für den Käufer ist, sind Verzögerungen äußerst nachteilig. Die Organisation, die nur begrenzte Möglichkeiten der Einflußnahme hat – außer dem Fabrikanten mit Nichterneuerung des Vertrages zu drohen –, befindet sich dann in einer sehr schwierigen Lage.

Ich betrachte die Notwendigkeit einer praktisch unfehlbaren inneren Organisation als den Hauptnachteil dieses Vorgehens.

Eine gute Organisation ist die Quintessenz der Industrialisierung. In einer Unternehmung, die nur eine technische und organisatorische Funktion ausübt, ist «Organisation» gleichbedeutend mit Lebensfähigkeit. Die geringste Störung kann katastrophale Ausmaße annehmen, welche das ganze Vorgehen in Frage stellen und unter Umständen das Bestehen der Firma bedrohen würden.

Zwischen den beiden oben beschriebenen Extremen gibt es natürlich viele Arten des Vorgehens, die je nach den örtlichen Umständen die Vorteile dieser oder jener Methoden zu kombinieren suchen.

Wir dürfen nicht vergessen, daß das Wesen der Industrialisierung nicht in einer bestimmten Produktionsweise, sondern vor allem in einer bestimmten Haltung zu suchen ist. Um eine Konstanz in der Produktion zu erreichen – mit anderen Worten, um fähig zu sein, das Lager auf angemessener Höhe zu halten und die optimale Leistung aus der Produktionsausrüstung herauszuholen –, ist eine Kontrolle des Absatzmarktes notwendig. Ohne diese Grundbedingungen handelt es sich nur um technologisch-fortschrittliches konventionelles Bauen, aber nie um Bauindustrialisierung.

client the advantages of a perfectly engineered product (through its technical development) unbiased by any material or production line, and of the cost-saving achieved by the intervention of an organization which can obtain through bulk-buying noticeable discounts. When the "client" is an architect the advantages of a pre-engineered building are evident: all detailing is solved. To the contractor reliable delivery and time-saving factors become convincing arguments.

That it is possible to succeed with this kind of organization has been proved, mainly in Western Europe, by the considerable success obtained in Germany by Brockhouse-Systembau; in France by Constructions Modulaires and in Spain by the Servicio Técnico de Construcciones Modulares. It may be of some interest to note that all these organizations operate within the field of educational or public facilities and that no example exists of similar ventures in the housing field.

From a qualitative angle the above-mentioned approach offers the main advantage of great flexibility. An organization operating under this scheme may introduce new sub-systems into its basic design practically at will. The only limitation may be the yearly or two-yearly contract with a specific manufacturer.

Obviously this organizational set-up also presents certain disadvantages. The most serious one may occur in connection with an unexpected marketing success. Then the selected manufacturers may prove incapable of delivering the components within the predetermined time-schedule.

Time-saving being a major selling-point, delays become serious drawbacks. The organization, having limited direct intervention-power (besides threatening the uncomplying manufacturer with a non-renewal of the contract) may then find itself in a very difficult position.

I see as the major drawback of this operational form the requirement for a practically infallible internal organization.

It will never be repeated enough that the whole essence of industrialization is good organization. In the case of a company assuming a merely technical and organizational rôle, "organization" becomes synonimous of viability. A minor breakdown can reach catastrophic dimensions, bringing into doubt the whole method and threatening at times the existence of the company.

Clearly between the two extreme conceptions exemplified above exist all kinds of mixed ventures, attempting to combine, according to local circumstances, the advantages of one or the other operational form.

Still it may be worthwhile to underline the fact that industrialization does not end in a certain form of production, but is mainly a result of a specific attitude. To achieve a reasonable consistency of production (which means in other terms to be capable of retaining the specially-trained workers and technicians, keeping stocks at a reasonable level, and obtaining the optimum of productivity from the equipment) it becomes necessary to control the marketing outlets. Without these basic conditions we can only talk of technologically-sophisticated conventional building, never of industrialized building.

4.3 Organisatorische Schwierigkeiten

Es mag von Nutzen sein, sich den Aufbau einer industrialisierten Bauorganisation einmal näher anzusehen. Zuvor müssen wir aber den Leser warnen: die folgenden Seiten sind mit äußerster Vorsicht zu lesen. Einige Prinzipien sind zweifellos immer gültig, aber Extrapolationen sind gefährlich. Auch können in vielen Fällen lokale Umstände eine Arbeitssituation so verändern, daß die ganze begriffliche Unterlage ins Wanken gerät.

Die Veranlassung, uns mit diesem Thema zu befassen, liegt in der einfachen Tatsache, daß scheinbar «selbstverständliche» Wahrheiten sehr oft vergessen werden und viele sonst ganz vernünftig geleiteten Unternehmen dadurch bankrott gegangen sind.

Wie schon vorher gesagt, muß der Unternehmer eine tadellose Innenorganisation haben, um erfolgreich zu sein. Niemand wird das bestreiten. Aber worin besteht eine fehlerfreie Innenorganisation? Wir wollen versuchen, das Problem dem theoretischen Aufbau einer komplexen Organisation gemäß zu analysieren. Wir können einige Hauptfunktionen feststellen:
a) Der Entscheidungsprozeß
b) Beschaffung von Informationen
c) Verarbeitung von Informationen
d) Produktion
e) Montage.

4.31 Der Entscheidungsprozeß

Der Entscheidungsprozeß in der Bauindustrie unterscheidet sich offensichtlich nicht vom Entscheidungsprozeß in anderen Gebieten. Hierüber gibt es sehr ernsthafte und trockene Veröffentlichungen, und alle Arten von Hilfsmitteln für das Management sind entwickelt worden.

Der Leser möge zur Erfrischung «Up the Organization» von Robert Townsend genießen. Wer Zweifel über seine eigene Organisation und seine dortige Stellung hat, dem könnte durch die Lektüre des «Peterprinzips» von Dr. Laurence J. Peter und des klassischen «The Parkinson Law» von G. Northcote Parkinson geholfen werden.

4.32 Beschaffung von Informationen

Die Lage ist anders, wenn es sich um das Beschaffen von Informationen handelt. Hier gibt es einen Unterschied zwischen der Bauindustrie und anderen Industrien. Wir können folgende Arten der erforderlichen Informationen feststellen:
a) Technische und architektonische Informationen
b) Marktinformationen
c) Finanzielle Informationen.

Die Beschaffung von technischen Informationen kann ein breites Gebiet umfassen: Design von Montagefolgen und Montagewerkzeugen, Design von Produktionswerkzeugen, Geräteanordnung und Produktionsablauf.

Testanlagen und Werkstätten für Prototypen sind notwendig. Lehrstätten mit Ausbildungskursen für technisches und Produktionspersonal könnten manche Ratlosigkeit auf der Baustelle gut vermeiden.

Im Bereich der Beschaffung von technischen Informationen trifft man oft eigenartige Haltungen. Viele Firmen verlassen sich immer noch auf konventionelle, statische Berechnungen für das Dimensionieren ihrer

4.3 Organizational difficulties

It may well be worthwhile to take a closer look at the set-up of an industrialized building organization. Before doing so we should, however, warn the reader to approach the following pages with the utmost caution. A few principles are certainly always valid, but extrapolations are dangerous. Also in most cases local factors may well influence operational situations to the extent of upsetting the whole intellectual framework.

The justification for involving ourselves with this subject is the simple fact that these apparently "self-evident" truths are very often forgotten and many otherwise – sensible industrialized ventures have gone bankrupt in the process.

We have already mentioned that to operate successfully, a correct internal organization is required. Let us try to analyse the problem following a theoretical build-up of a complex organization. We can identify a few major functions:
a) Decision making
b) Production of information
c) Processing of information
d) Production
e) Assembly.

4.31 Decision making

Decision making in the building industry is obviously not different from decison making in any other field. Very serious theories are available and all kinds of management-tools have been developed.

However, the reader may well read as a refreshing experience *Up the Organization* by Robert Townsend. Helpful for the reader in doubt about his own organization and his position within it may be *The Peter Principle* by Dr. Laurence J. Peter and the classic *The Parkinson Law* by G. Northcote Parkinson.

4.32 Production of information

The situation changes when we concern ourselves with the production of information. Here the situation of the building industry differs slightly from that of the other industries. We can identify the following kind of information which must be produced:
a) Technical and architectural information
b) Market information
c) Financial information.

The production of technical information may well cover a very wide area: Design of assembly-sequences and assembly-tools, design of production-tools, production-lines and production-sequences.

Testing facilities and workshops for full-scale models become necessary. Teaching laboratories to instruct the production and technical staff may well avoid the on-site head-scratching so common today.

In this area of the production of technical information the most foolish attitudes can be found. Many companies continue to rely upon conventional structural calculations for the design of their components. Now a big advantage of industrialized building lies in the

Bauelemente. Ein großer Vorteil des industrialisierten Bauens liegt aber gerade in der Massenproduktion gewisser Teile unter kontrollierbaren Verhältnissen. Statische Berechnungen ergeben eine annähernde, aber sichere Dimensionierung. Sind die Produktions- und Montageabläufe unter strenger Kontrolle, können mit korrekten Tests erhebliche Ersparnisse erzielt werden. Dazu sind aber gründliche Kenntnisse erforderlich, und entsprechende Maßnahmen müssen getroffen werden, um zu garantieren, daß auf dem Bauplatz gemäß den Montageanleitungen montiert wird.

Zu diesem Zweck müssen Seminare durchgeführt werden, um dem Aufsichtspersonal das neue Bauprojekt, seine Vorteile und möglichen Schwächen zu erklären.

Hier kann man bei den meisten kommerziellen Organisationen ein merkwürdiges Phänomen feststellen: Die gleiche Firma, die ohne zu zögern über zweihunderttausend Dollar für eine neue Maschine ausgibt, ist gleichzeitig nicht geneigt, ihrem Bauleiter ein dreitägiges Seminar zu vermitteln oder das Produkt einem Zerstörungstest zu unterwerfen. Die Firma, die eine solche Haltung einnimmt, schadet sich selbst, obwohl sie einer eingefahrenen Tradition folgt. Irgendwann wird man aus einigen Metallteilen ein fertiges Produkt bauen müssen. Es hilft, wenn Vorgang und Resultat zum vornherein bekannt sind.

Bei der Dimensionierung kann eine einfache Regel angewendet werden: nie berechnen, immer testen.

Die Beschaffung von architektonischen Informationen muß sich den verfügbaren Herstellungsmethoden anpassen. Eine Architekturabteilung oder ein assoziierter Architekt, der ständig Entwürfe liefert, die angepaßt werden müssen, ist bestenfalls eine Last. Dadurch entstehen Verzögerungen in der Produktion und Montage und erhebliche Extrakosten.

In einer «idealen» Organisation gibt es eine Architekturabteilung, eine Entwicklungsabteilung, eine statische Abteilung und eine Produktionsabteilung, die alle technischen Informationen beschaffen. Die Entwicklungsabteilung wird in eine langfristige und eine projektgebundene Entwicklungsgruppe eingeteilt. Oft unterschätzen Organisationen, für die Bauindustrialisierung ein neues Gebiet ist, die Qualität der erforderlichen Ad-hoc-Entwicklungen und überlassen es dem Bauleiter, eine Lösung zu finden; dann gibt es gewöhnlich Überraschungen. Auch die Abteilung für Marktforschung ist an der Beschaffung von Informationen beteiligt. Obgleich ihre Funktion für einen erfolgreichen Arbeitsbetrieb immer wesentlich ist, gibt es erhebliche Unterschiede in der Gründlichkeit und im Umfang ihrer Tätigkeit. In der einfachsten Form wendet sie sich an potentielle Auftraggeber und liefert Informationen über die Absatzmöglichkeiten des Produktes an die Geschäftsleitung. Einige Organisationen treiben eine eigentliche Marktforschung, um die Chancen eines entworfenen, aber noch nicht hergestellten Produktes abzuschätzen.

Nur wenige Organisationen gehen so weit, die Bedürfnisse potentieller Benutzer ernsthaft zu erforschen, um die zukünftigen Eigenschaften ihres Produktes bestimmen zu können.

In den meisten Organisationen läuft parallel zur Marktforschung die ihr verwandte Tätigkeit der Werbung und der Public Relations. Die Analysen neuer Produkte

mass production of certain items under controllable conditions. Structural calculations give (and rightly so) dimensions which can only be approximate but very safe. If we can be sure of our production and assembly, considerable saving can be achieved by correct tests. However, a deep knowledge is necessary and suitable measures must be taken to insure that on the assembly site the correct assembly conditions are recreated.

To achieve this, proper educational seminars must be given to all supervising personnel explaining the new design, its advantages and its critical features.

Somehow a strange phenomenon takes place at this stage in most commercial organizations: the same management who without a shadow of doubt willingly invests some two hundred thousand dollars in a new piece of equipment becomes very doubtful about giving three-day seminars to his supervising staff or in allowing a few destructive tests to be carried out. This management, while certainly following a well-grounded tradition, does a great disservice to its organization. Along the line people will be handed a few bits of metal to transform into a product.

Knowing what and how they should do it usually helps.

On the design activity a simple rule may be applied: never calculate, always test.

The production of architectural information has to be strictly related to the production methods available. An architectural department, or an associated architect, who constantly turns out designs which require adaptations is at best a nuisance. Normally delays in production and assembly, and considerable extra costs, will be caused.

Within an "ideal" organization we may find a department of architecture, a development department, a production department and a structure department involved in the production of technical information, whereby the development department may be broken down in long range and "running" development (often organizations new at the game of industrialized building tend to underestimate the quantity of "ad hoc" development required and leave it to the site-supervisor to find a solution, usually with surprising results). Also involved in the production of information we have the marketing department. The activity of the marketing department, while always essential, varies in depth and amplitude considerably. In its simplest form the marketing department merely approaches potential clients and provides information on the marketing ability of the product. (A few organizations have gone further and have involved themselves in proper market-research in order to assess the chances for success of a product as yet on the drawing board.) Only few organizations have started serious research into the needs of the potential users, in order to determine the future characteristics of their product.

Parallel to the marketing activity we find in most organizations the connected advertising and public-relation department. We can also consider as part of the production of market information the activity involving the analysis of new products and production procedures. This kind of information is essential for a correct design and only too often this function is completely neglected and left to the hazards of magazine reading and salesmen visits. Collecting informa-

und Herstellungsmethoden können auch zur Bestimmung des allgemeinen Marktbildes beitragen. Die Beschaffung dieser Informationen ist wesentlich für ein korrektes Design, sie wird oft vernachlässigt und den Zufälligkeiten des Lesens von Zeitschriften und der Besuche von Geschäftsreisenden überlassen. Die Beschaffung von Informationen über erhältliche Produkte und Kostentrends kann eine sehr umfassende Tätigkeit sein. In ihrer höchsten Form grenzt sie an industrielle Spionage, und die einfachste Form bedeutet weiter nichts als das Lesen lokaler Veröffentlichungen der Kosten- und Lohntarife.

Die Beschaffung von finanziellen Informationen unterrichtet das Management über die genaue finanzielle Situation der Firma und liefert Voraussagen auf mittelfristige Sicht über die Finanzlage des Marktes. Da die Bautätigkeit ständig Konjunkturschwankungen ausgesetzt ist, bildet die korrekte Voraussage der Finanzbewegungen einen wichtigen Anhaltspunkt für Einschränkung oder Ausdehnung der Produktion, des Lagers, der Werbung usw.

4.33 Verarbeitung von Informationen

Während die Beschaffung von Informationen im allgemeinen ähnlich ist wie in anderen Industrien, so hat die Verarbeitung von Informationen in einer industrialisierten Baufirma einen eigenen Charakter.
Der Grund hierfür kann in den Besonderheiten des Marktes gefunden werden. Wir können annehmen, daß die von uns analysierte «ideale» Firma sowohl mit ihrer eigenen Architekturabteilung als auch mit auswärtigen Architekten arbeitet. In beiden Fällen paßt sich der Entwurf den Bedürfnissen des Bauherrn an. Vor Beginn der Produktion muß der fertige Entwurf von einer Reihe von Abteilungen innerhalb der Organisation geprüft werden. Gewöhnlich sind folgende Punkte zu beachten:

a) Prüfung, ob der Entwurf innerhalb der Produktionsmöglichkeiten der Firma liegt (siehe auch 4.4)
b) Zerlegung des Entwurfes in seine Produktionskomponenten
c) Entwurf der speziellen Komponenten, die erforder-

tion on available products and on cost-trends can be a very involving activity. At its highest level it touches on industrial espionage and in its simplest form simply means the survey of local cost and wages magazines.

The production of financial information has as objective to keep the management informed on the correct financial situation of the company, and attempts to draw medium-range forecasts on the situation of the financial market.
The building activity being in fact recurrently prone to stop-and-go practices, a correct forecast of financial movements is a very useful guide for the restriction or expansion of the production rate, stocks, advertising, etc.

4.33 Processing of information

While the production of information is, in general terms, common to other industries, the processing of information in the industrialized building company is typical of this branch of activity.
The reason for this situation can be found in the peculiarity of the market. We can assume that the "ideal" company under analysis is working both with its own architectural department and with "outside" architects. In both cases the design offered by the architects will have been influenced by the client's needs and requirements. This design produced shall have to be vetted by a series of agencies within the organization, prior to production.
The usual steps required are:

a) Check if the design can be produced within the standards of the company (see also 4.4)
b) Check the structural safety of the design
c) Break down the design into its production components

Das Funktionsschema einer führenden britischen Firma (mit Erlaubnis von Vic Hallam Ltd.)

1	Auftraggeber	30	Behandlung des Materials
2	Verkaufsplanung	31	Terminplan für Baustoffe
3	Werbung	32	Zeichnungsbüro
4	Vertrieb	33	Lager
5	Public Relations	34	Einkauf
6	Marktforschung	35	Holzeinkauf
7	Einschätzung	36	Brennerei
8	Herstellungsnormen	37	Herstellung
9	Forschung und Entwicklung	38	Baukontrolle
10	Entwicklungskoordination	39	Inspektion und Unterhalt
11	Entwurf des Produktes	40	Einbau und Ausstattung
12	Kostenplanung	41	Glasarbeiten
13	Vorausplanung	42	Dacharbeiten
14	Finanzkontrolle	43	Decken- und Bodenarbeiten
15	Produktionsplanung	44	Malerarbeiten und Innendekoration
16	Kostenberechnung	45	Elektrische Installationen
17	Arbeitsplan	46	Sanitäre Installationen
18	Statische Berechnungen	47	Heizung
19	Prüfung und Inspektion der Komponenten	48	Bau
20	Balkenfabrikation	49	Transport
21	Lager der Komponenten	50	Verkehrskontrolle
22	Produktionsvorgang	51	Inspektion des Unterbaus
23	Malerarbeiten	52	Generalunternehmer
24	Montage	53	Auswärtige Arbeiten und Dienste
25	Lagerung der Subkomponenten	54	Fundamente und Platten
26	Werkzeuge für Maschinen	55	Spezielle Außen- und Innenwandfertigung
27	Formgebung	56	Abgeschlossener Vertrag
28	Einstellung der Einrichtung		
29	Fräserei		

The functional scheme of a leading British company (courtesy of Vic Hallam Ltd.)

1	Client	29	Milling
2	Sales planning	30	Material handling
3	Publicity	31	Material scheduling
4	Sales promotion	32	Work drawing office
5	Public relations	33	Stores
6	Market research	34	Purchasing
7	Estimating	35	Timber buying
8	Manufacturing standards	36	Kilning
9	Research and development	37	Manufacture
10	Development co-ordination	38	Building control
11	Product design	39	Inspection and maintenance
12	Cost planning	40	Fitting and furnishing
13	Forecasting	41	Glazing
14	Financial control	42	Roofing
15	Production planning	43	Floors
16	Costing	44	Painting and decorating
17	Work study	45	Electrical
18	Engineering	46	Plumbing
19	Component testing and inspection	47	Heating
20	Beam fabrication	48	Erection
21	Component store	49	Transport
22	Phasing	50	Traffic control
23	Painting	51	Substructure inspection
24	Assembly	52	General contractor
25	Sub-component storage	53	External works and services
26	Machine tooling	54	Foundations and site slab
27	Machining	55	Special internal and external wall finishes
28	Setting out	56	Completed contract

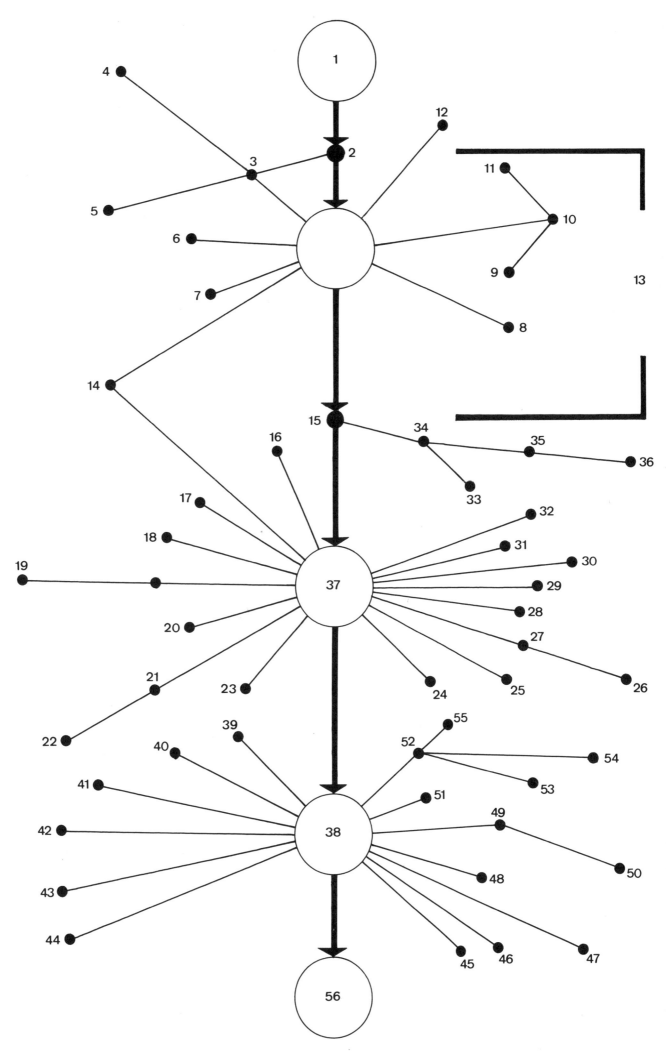

lich sind und/oder der lokalen Subsysteme: Fundament, Erdbewegung usw.
d) Unterteilung der Komponenten in ihre Elemente und Zubehör
e) Kalkulation der Gesamtbaukosten
f) Arbeitsplan für Produktion und Montage
g) Anfangssignal an Produktions- und Montageabteilungen
h) Instruktionen über das neue Projekt an die Bauleitung
i) Kontakt mit Lokalunternehmen für kleinere Aufgaben, zum Beispiel Gärtner- und Malerarbeiten usw.
Natürlich ändert sich diese Einteilung von einer Organisation zur andern, von einem Lande zum anderen; lokale Bedingungen sind von großer Wichtigkeit.
Alle oben erwähnten Phasen sind nötig, um einen reibungslosen Produktions- und Montageablauf sicherzustellen und die vorhandenen Mittel richtig zu plazieren. Wenn die Arbeit läuft, muß ein ständiger Strom der Informationen von der Werkhalle und Baustelle der Geschäftsleitung ermöglichen, die erforderlichen Maßnahmen zu treffen, um Terminplan und Kostenniveau planmäßig einzuhalten. Die Innenorganisation des Informationsflusses nimmt von einer Organisation zur andern viele verschiedene Formen an, aber glücklicherweise ist auf diesem Gebiete ein großes Fachwissen vorhanden. Welche Methode auch immer angewandt wird, die Leitung muß in ständigem Kontakt mit dem Arbeitsvorgang sein. Nur zu oft realisierte die Geschäftsleitung erst nach Abschluß des Projektes, daß der Terminplan nicht einzuhalten war oder das Budget überschritten wurde. Aber dann ist es zu spät für ein wirksames Eingreifen.

4.34 Produktion
In der Werkhalle stößt die industrialisierte Bauorganisation auf dieselben Probleme, die auch für andere Industriezweige typisch sind.
Auch hier kann man wieder zwei Hauptgebiete unterscheiden:
a) Technische Probleme
b) Wirtschaftlich-technologische Probleme.
Technische Probleme können gewöhnlich ziemlich leicht gelöst werden. Qualitätskontrolle, Umgang mit dem Produkt, Toleranzen, diese Dinge können in der Regel von qualifizierten Technikern gemeistert werden. Ein Spezialfall ist die Firma, die Komponenten nicht selber herstellt, sondern von anderen Fabrikanten anfertigen läßt. Das Problem wird akut, wenn der erforderte Qualitätsstandard des Produkts die üblichen Normen der Lieferfirma übersteigt. In dem Falle wird Druck von seiten der Geschäftsleitung und geeignete Schulung der Betriebstechniker erforderlich. Sehr oft muß eine ständige Qualitätskontrolle durch einen taktvollen und fähigen Techniker durchgeführt werden. Der Arbeitsablauf der Komponentenhersteller muß für den Erfolg des Gesamtunternehmens reibungslos funktionieren, und seine Fähigkeiten und guten Absichten allein genügen nicht; er hat ja noch andere Sorgen.
Das Thema wird schwieriger, wenn es sich um die Wirtschaftlichkeit und Technologie der Produktion handelt. Außer dem selbstverständlichen Ziel, die Kosten niedrig zu halten, erhebt sich das schwierige Problem des Vorratslagers. Es ist nicht einfach, eine Projekt-Produktion aufrechtzuerhalten. Die Bautätig-

d) Design the special components necessary and/or the local sub-systems: foundations, earth-movements, etc.
e) Break down the components into their elements and bits and pieces
f) Recalculate the cost of the complete building
g) Schedule production and assembly
h) Transmit production and assembly "go" to the relevant units
i) Instruct site-supervision with the new job
k) Contact local operator for minor operations (gardening, painting, etc.).
Naturally the above break down will vary from organization to organization and from country to country (local habits definitely play a relevant rôle).
All the above mentioned stages are necessary in order to insure smooth production and assembly process and to allocate correctly the available resources. When the job is under way a constant flow of information from the factory-floor and the building-site is required to enable the management to take the necessary measures for maintaining the job on its schedule and on its planned cost-platform. The internal organization of the information flow varies considerably from organization to organization, but fortunately in this field considerable expertise is available.
Whichever method is followed the management must keep in touch with the job. Too often only after completion do the managers realise that the job has cost too much or that delays have taken place. But then it is too late for any useful intervention.

4.34 Production
On the factory-floor the industrialized building organization will find the typical problems of the other branches of industry.
Here again we can break down the subject into two major areas:
a) Technical problems
b) Economical-technological problems.
Technical problems can usually be solved with little difficulty. Quality control, handling of the product, tolerances, are all subjects which well-qualified technicians can usually master. A special case is represented by the company that does not produce the components, but relies on outside manufacturers.
The problem becomes acute when the product required exceeds the usual quality-standards of the manufacturers. Pressure at management level and qualified education of the factory-floor technicians becomes necessary. Very often constant control of the quality standards by a diplomatic and qualified technician will be required. The manufacturer of the components must operate smoothly for the success of the venture, and to rely simply on the ability and good will of the manufacturer may often prove insufficient: after all he has other problems to deal with!
The subject becomes more difficult when we concern ourselves with the economy and technology of the production. Besides the obvious aim of achieving a low-cost we are confronted with the tricky subject of stocks. It is very difficult indeed to maintain a job-oriented production. Building activity follows usually

keit hängt stark von der Saison ab, und nicht nur wettermäßig: Budgets werden meist im Frühjahr bewilligt; Schulen fangen im Herbst an; Feriensiedlungen und Hotels wollen bestimmte Eröffnungstermine einhalten usw. Die Projekt-Produktion wird dadurch erschwert und kann die Absatzmöglichkeiten der Firma erheblich einschränken. Das Vorratslager hingegen verursacht starke finanzielle Belastungen; einige Hersteller von Komponenten mußten dies am eigenen Leib spüren: die Firma ging bankrott. Das Vorratslager ist auch schwer aufzubauen. Wenn Betonpaneele gefertigt werden sollen, welche Größe, welche Farbe soll produziert werden? Manchmal können nur höchstverfeinerte statische Hilfsmittel der Geschäftsleitung helfen, die Lage zu meistern. Oder man versucht den Markt selbst zu beherrschen.

Kopfzerbrechen bereitet auch die Anschaffung des Maschinenparkes. Die heutige technologische Entwicklung versieht uns mit hochspezialisierten Werkzeugen: Bohrautomaten, die nur in 32-mm-Abstand bohren; Schaumspritzgeräte, die nur in Verbindung mit einer spezifischen Form benutzt werden können usw. Ein spezifisches Gerät kann demnach nur *eine* bestimmte Aufgabe ausgezeichnet erfüllen. Aber gleichzeitig wirken diese Geräte bei künftigen Entwicklungen behindernd. Die Funktionsfähigkeit des Geräts muß bei neuen Produkten berücksichtigt werden, damit nicht schwere finanzielle Verluste folgen.

Dieses Problem spielt eine große Rolle in «baustofforientierten» Organisationen (siehe 4.2). Nach wenigen Jahren sind sie überladen mit Spezialgeräten und Fertigkeiten und können kaum der raschen Entwicklung des Marktes folgen. Die Alternative, nämlich die Benutzung sehr flexibler Maschinen, wie zum Beispiel verstellbare Formen oder allgemein verwendbare Kraftwerkzeuge, ist auch nicht immer zufriedenstellend. Bei allen Werkzeugen, die ein breites Anwendungsgebiet haben, stellt sich meistens heraus, daß ihre Flexibilität nie voll ausgenützt wird, daß sie die Arbeitszeit verlängern und damit die Kosten erhöhen. Das gleiche gilt natürlich auch für die Baustelle. Man muß die richtigen Geräte gebrauchen, aber manchmal ist es schwer zu bestimmen, welches die richtigen sind.

Die Ausbildung des Personals ist eine andere Seite des Produktionsproblems. Es wird allgemein behauptet, daß angelernte Arbeiter für industrialisierte Bauarbeiten verwendbar sind. Das ist zum Teil richtig, aber trotzdem müssen diese Arbeiter für bestimmte Aufgaben geschult werden, und nach der Schulung sollten sie auf jeden Fall in der gleichen Organisation gehalten werden. So entsteht wieder die Notwendigkeit, für das Lager zu produzieren, um den Arbeitern Beschäftigung zu sichern. Spezifische Fertigkeiten selbst für Aufgaben, die von den meisten Arbeitern geleistet werden können, bilden den Unterschied zwischen einem guten und einem schlechten Produkt (Fertigstellung eines Betonpaneels, Verglasen eines Fensters usw.). Wenn man die Auswechselbarkeit der Arbeiter und die Funktionsmöglichkeiten der Werkzeuge als selbstverständlich voraussetzt, dann steuert man einem Mißerfolg entgegen.

4.35 Montage
Montage auf dem Bauplatz ist die Tätigkeit, welche die Bauindustrie von allen andern unterscheidet. Kaum

a seasonable pattern (not only weather conditioned: budgets are approved usually in spring, schools open in autumn, holiday-housing and hotels aim at definite opening times, etc.). Job-production then becomes difficult or may limit considerably the market of the organization. Stocks however cause high financial costs (a few manufacturers of components have learned the lesson the hard way: through bankruptcy). Stocks also may be very difficult to organize. If the production of concrete panels is our objective, which size, which colour of panels should we produce? At times only the most sophisticated statistical management-tools can cope with the problem. (Or we may choose to avoid the problem by controlling our own market.)

Another headache is the acquisition of equipment. Our technological development tends to give us very specialized equipment: automatic drills which will drill at 32 mm interval only, foam-spraying equipment which can be used only in connection with a specific mould, etc. Obviously a specific piece of equipment can perform beautifully a certain duty but nothing else. The consequence is that it becomes a limiting factor for future developments. All new design will have to take into consideration the capabilities of the equipment or heavy financial penalties will be paid. The problem is very conspicuous in "material-oriented" organizations (see 4.2).

After a few years of operation they are cluttered with very specialized equipment and skills and can hardly follow the evolution of the market. The alternative of using very flexible equipment (adjustable moulds, general-use power-tools) is not always satisfactory either.

As with all tools designed to do a wide range of jobs it may prove that flexible equipment is never optimally utilized, is slow in operating and as a result, expensive.

Clearly the same rules apply on the building-site: the correct kind of equipment must be utilized, but at times the identification of the correct equipment is not an easy task.

The training of the staff is another facet of the production problem. It is commonly said that semi-skilled labour can be used in industrialized building. This is partially correct, but the problem remains that this labour needs to be trained to perform a specific task, and when trained it had better stay with the organization (and here again may emerge the necessity to produce for stocks in order to keep the people occupied). The possession of certain specific skills make all the difference between a good and a bad product (finishing a concrete panel, glazing of a window, etc.). Taking for granted the interchangeability of labour and the capability of the equipment usually leads to disaster.

4.35 Assembly
The assembly on site is the most specific activity of the building industry. Few other industries are sub-

eine andere Industrie ist einem solchen Druck ausgesetzt, wie es der ständige Wechsel des Arbeitsplatzes hier mit sich bringt. Theoretisch sollte die Bauindustrialisierung den Druck erleichtern. Dieses Ziel ist wirklich erreicht, wenn die Organisation reibungslos funktioniert und das Personal gut geschult ist. Dieser Idealzustand ist aber nur schwer erreichbar.

Das Wetter kann immer einen Strich durch die Rechnung des bestgeplanten Programmes machen, und eine verspätete – oder verfrühte – Lieferung kann erhebliche Schwierigkeiten bereiten. In gewisser Weise wird industrialisiertes Bauen mehr von verspäteten Lieferungen beeinträchtigt als konventionelles Bauen; die Montagefolge ist in der Regel weniger flexibel. Auch ist eine strengere Qualitätskontrolle erforderlich.

Durch genaues Dimensionieren der Bauelemente werden große Ersparnisse erzielt. Wir haben auch bereits erwähnt, daß angelernte Arbeiter an der Baustelle beschäftigt werden können. Beide Faktoren zeigen deutlich, daß hochqualifizierte Aufsicht nötig ist. Die Wichtigkeit des Postens eines Bauleiters wird leider zu oft unterschätzt. Dieser Mann hat die gleiche Schlüsselstellung im Produktionsprozeß wie der Einkäufer im Planungsprozeß. Der Bauleiter muß die besonderen Merkmale seines Produkts in- und auswendig kennen und fähig sein, Arbeitsvorgänge und Qualitätsnormen seinen oft widerstrebenden Untergebenen aufzuzwingen. Er muß die Geschäftsleitung von drohenden Schwierigkeiten unterrichten und nicht warten, bis die Schwierigkeiten wirklich eintreten, um dann verzweifelt zur nächsten Telefonkabine zu laufen.

Ich habe einen guten Bauleiter gesehen, der mit absolut minderwertigen Werkzeugen eine Aufgabe erfolgreich zu Ende führte, während die besten Geräte unter einem unfähigen Mann nutzlos herumstehen.

Zusammenfassend möchte ich betonen, daß bei der Produktion und Montage industrialisierter Bauten der Auswahl und der Schulung des Personals viel mehr Wichtigkeit beigemessen werden sollte als der Anschaffung hochmoderner Bauwerkzeuge. Solche Geräte sollten nur gekauft werden, wenn keine andere Lösung offensteht, und auch nur dann, wenn die Marktvoraussage sehr optimistisch ist (mit einem Sicherheitsfaktor von 99%). Der Ankauf von Bauwerkzeugen, um die Ambitionen der Generaldirektoren und Aufsichtsräte zu befriedigen, und die Tendenz, die menschlichen Stützen außer acht zu lassen, hat mehr Fehlschläge verursacht, als man zugeben will.

4.4 Operationelle Taktiken

Das Hauptziel der industrialisierten Bauorganisation besteht darin, den höchsten Grad des Erfolges zu erreichen, ganz gleich welche organisatorischen Vorgehen gewählt werden. Dieser Grad des Erfolges kann auf denkbar viele Weisen gemessen werden: durch Einkommen, persönliches Prestige oder die Befriedigung eines bestehenden Bedarfs, aber in der Praxis kann er immer an der leicht meßbaren Anzahl der errichteten Gebäude beurteilt werden. Jedoch können wir von diesen Zielen nicht unmittelbar den Weg zum Erfolg ableiten. Aber wir müssen uns bewußt sein, daß die meisten Organisationen nicht wissen,

---

jected to the stresses caused by the constant change of operating site. Theoretically the industrialization of building should reduce such stresses. The aim is indeed achieved when the organization is running smoothly and the staff is properly trained. However, this happy state of affairs is not easily achieved.

The weather can still upset a well-planned operation, and a delayed (or anticipated) delivery can cause considerable difficulties. To some extent industrialized buildings are more affected by a delayed delivery than conventional buildings (the assembly sequence is usually less flexible).

Also a tighter quality-control is required. We have said before that considerable savings can be achieved by the exact dimensioning of the components. We have also mentioned that semi-skilled labour can be employed on an industrialized building-site: both factors, however, imply the need of a very capable supervision. The importance of the site-supervisor's job is unfortunately too often understimated. On the production line this man occupies the same key position that the acquisition man occupies in the design line. The site-supervisor must know the specific features of his product by heart and be capable of enforcing procedures and quality standards onto his usual reluctant subordinates. He has to inform the headquarters about impending difficulties and not run frantically for the next phone-box when the difficulties actually occur.

I have seen a good site-supervisor successfully complete a job with the most appalling equipment while the best equipment has been left idle under an incompetent man.

To summarize the production and assembly of industrialized building I would always suggest that much more importance should be given to the selection and training of the staff than to the purchase of sophisticated equipment. Also sophisticated equipment should be bought only when no other solution is available, and only when a very optimistic (with some 99% safety factor) market forecast is available. The acquisition of equipment to satisfy the egos of managing directors and boards and the tendancy to forget about the human props has caused more failures than people are willing to admit.

4.4 Operational tactics

Under whichever kind of organizational form, the primary aim of the industrialized building organization is to achieve the highest possible degree of success. This degree of success may be measured in many different ways: through income, personal prestige or the satisfaction of an existing need, but in practice it can always be translated in the easily measured quantity of erected buildings. However, from the very simple aims mentioned above we cannot immediately obtain a simple path towards success. In fact we have to recognize that most organizations are ignorant of the aim they are pursuing. This is a common

welches Ziel sie verfolgen. Diese Schwäche wird zwar von vielen anderen Industrien geteilt, aber das macht sie nicht weniger peinlich. Wir kennen Organisationen, deren Ziel anscheinend ein größerer Ertrag auf ihr investiertes Kapital ist. Eine genauere Analyse ergibt aber, daß ihr wahres Ziel der narzißtische Wunsch ist, sich ein «Reich» zu bauen. So werden sie in Übersee-Unternehmungen verwickelt oder in der Einrichtung von Produktionsanlagen, immer unter der Flagge des gesunden Menschenverstandes und der Geschäftstüchtigkeit, aber in Wirklichkeit für das Prestige und die persönliche Genugtuung der Geschäftsleitung.

Alle Organisationen brauchen, um erfolgreich zu sein, eine Strategie auf lange Sicht und kurzfristige «Taktiken». Gewöhnlich jedoch werden nur die Taktiken definiert – meistens unter Druck von außen –, und die Strategie bleibt sich selbst überlassen. Im Ergebnis verfolgt die Organisation einen ständig wechselnden Kurs und bringt ihr Personal und oft auch ihre Kunden zur Verzweiflung. Tatsächlich stehen einer industrialisierten Organisation alle möglichen Strategien zur Verfügung, und alle können unter gewissen Voraussetzungen zufriedenstellen sein. Die Strategie, die auf der Hand zu liegen scheint, ist eine größere Beteiligung am Absatzmarkt und der Versuch, sich interessante Aufträge zu sichern. Eine andere übliche ist eine Erweiterung des Tätigkeitsfeldes durch die Fertigung von Komponenten, die andern Unternehmern zum Verkauf angeboten werden.

Der Verkauf von «Know-how»-Lizenzen an Dritte ist eine weitere interessante Möglichkeit. In diesem Falle dient die eigene Produktion als nützliches Schaufenster für etwaige Käufer. Auch können Dienstleistungen technischer und finanzieller Art an Konkurrenten verkauft werden; dann wird die technische oder Finanzabteilung der Organisation wichtiger als alle anderen Abteilungen.

Es ist auch möglich, daß eine Organisation sich mit industrialisiertem Bauen abgibt einfach als Schulungsprozeß für ihr vorhandenes Personal, das sich mit konventionellen und vielleicht überholten Tätigkeiten befaßt. Hier mögen einige Beispiele helfen.

Wenn wir annehmen, daß eine Bank und ein Stahlbetrieb sich für die Gründung einer industrialisierten Bauorganisation zum Zwecke des Wohnungsbaus zusammentun, so sollten wir wohl voraussetzen, daß ihr Hauptzweck die Produktion von Wohnungen sei. In Wirklichkeit aber ist der Bau von Wohnhäusern nur ein Nebenzweck für beide Teilhaber. Der Bankier ist daran interessiert, Geld zu einem guten Zinssatz mit minimalem Risiko auszuleihen. Darum dünkt ihn die Bautätigkeit ein gutes Gebiet für sein Hauptinteresse, und man kann kaum von ihm erwarten, daß er sich interessiert um das Endprodukt kümmert. Für den Stahlfabrikanten, dem eine antiquierte Stahlbaufabrik gehört, ist der Hauptzweck, neue Techniken und Technologien in seinem Betrieb einzuführen. Er betrachtet die neue Organisation als eine Quelle frischer Ideen und möglicherweise als Kunden für seine Produkte. Auch hier kommt der direkte Erfolg der Gesellschaft erst in zweiter Linie.

Ein andersartiger Fall ist das Beispiel einer Organisation, die im eigenen Lande ein industrialisiertes Bausystem vertreibt. Aus allen möglichen Gründen ist die Gesellschaft am Auslandmarkt interessiert. Direkter Export des Produktes ist jedoch nicht möglich wegen

enough weakness in many industries but a very embarassing one. We have found organizations who apparently aim at a higher return of their investment. Deeper analysis shows, however, that their real aim is the narcissistic desire for "empire building". So they become involved in activities abroad, or in the production of unprofitable facilities, always under the flag of common sense and business acumen, but really for the prestige or personal satisfaction of the managers. All organizations, in order to operate successfully, need a long-term strategy and short-term "tactics". Usually, however, only the tactics are defined and the strategy is left alone to take care of itself. The result is that the organization follows a continually changing course, to the despair of the staff and the client. Indeed all kinds of long-term strategies are available to an industrialized organization, and all of them can be equally satisfactory under a given set of circumstances. The obvious one is to increase their share in the market, attempting to secure interesting jobs. Another common one is to broaden the area of activity by indulging in the production of components for sale to other operators in the field.

Also interesting may be the sale of know-how licenses to third parties in which case home-production becomes a simple display-window for possible buyers.

Services of technical or financial nature can also be sold to competitors in which case the importance of the technical or financial department will become paramount to the other activities of the organization.

It may also be possible for an organization to get involved in industrialized building simply to educate their staff, which is involved in more conventional and possibly outdated activities. Some examples may be of help on this subject.

If we assume that a bank and a steel manufacturer unite to form an industrialized building organization to produce housing, we may well believe that their primary objective is the production of dwellings. In actual fact the production of dwellings is only a secondary objective for the two partners. The banker is interested in the business of lending money at a good return and with the minimum risk. He therefore finds in the activity of building an interesting area for exercising his primary interest and he can hardly be expected to care much for the final product.

For the steel manufacturer, owning an antiquated steel-structure factory, the main objective is to introduce new techniques and technologies into his organization. He looks upon the new company as upon a source for new ideas, and possibly as a client for his own product. Again, in his case, the direct success of the company is only of secondary interest.

A different case may be exemplified by an organization which in its own country operates a certain industrialized building system. The organization is, for all kinds of reasons, interested in the foreign markets. However, direct export of the product is excluded because of the import restrictions enforced by the foreign government(s). The possible solutions lie in the creation of a fully-owned local subsidiary, in a partnership with some local operators, or in the straight sale of the "know-how". At this point the organization discovers that its own product is technically unacceptable to the local market (because of

der ausländischen Importbeschränkungen. Ein Ausweg ist die Gründung einer ihr voll gehörenden Tochtergesellschaft am Ort in Partnerschaft mit einem dortigen Unternehmer oder durch den direkten Verkauf des «Know-how». Aber dann entdeckt die Organisation, daß ihr eigenes Produkt für den dortigen Markt technisch unannehmbar ist (wegen Vorschriften, Kostengrenzen, Mangel an Baustoffen usw.). Offensichtlich ist die einzige wirklich verkäufliche Ware der Firma ihre Kenntnis der Organisation und der Verwaltungstechnik. Gewöhnlich aber wird eine Abmachung getroffen, in der die Originalfirma ihr technisches «Know-how» an die neue Organisation verkauft — selbst wenn die neue Organisation volles Eigentum der Originalfirma ist —, während die Geschäftsleiter ortsansässig sind. Erst nach Jahren ernster Schwierigkeiten realisiert die Mutterfirma, daß sie ein nicht lohnendes Produkt — die Technik — verkauft hat, und wenn noch Zeit ist, wird dann der Versuch gemacht, mit verwaltungstechnischen Kenntnissen auszuhelfen. Es ist merkwürdig, daß industrielle Organisationen über ihr wertvollstes Produkt in Unkenntnis sind, aber eine sorgfältige Analyse zeigt, daß solches die Regel ist und nicht die Ausnahme.

Nur wenig kann getan werden, um lohnenswerte Strategien zu definieren: jeder Fall muß für sich behandelt werden, und Verallgemeinerungen sind nicht möglich.

Taktiken sind leichter zu handhaben. Wiederum kann nicht jede Organisation den gleichen Weg verfolgen, aber einige Grundregeln sollen hier doch skizziert werden.

In normalen Verhältnissen ist der Idealzustand einer industrialisierten Baufirma der einer absoluten Beherrschung ihres eigenen Marktes. Diese Tatsache wurde kürzlich von Masayumi Yokoyama einem spanischen Kunden gegenüber wie folgt formuliert: «In einem industrialisierten Prozeß tauscht die Bauorganisation die passive Rolle des Lieferns von Diensten gegen die aktivere Rolle des Produzierens aus.»

Wir müssen uns aber klarmachen, daß der Weg zur Beherrschung des Marktes nicht leicht ist und jedenfalls erst nach einer langen Entwicklungsperiode gelingt. Die meisten Organisationen müssen deshalb mit den bestehenden Umständen vorlieb nehmen, in deren Rahmen man mehrere Teilnehmer feststellen kann:

a) Der Käufer (die Person oder Behörde, die das Gebäude kauft)
b) Der Benutzer (die Person oder Behörde, die das Gebäude benutzt)
c) Der Bauunternehmer (die Organisation, die das Gebäude montiert und die manchmal gleichzeitig die industrialisierte Organisation sein kann)
d) Die öffentliche Behörde (die Baubewilligungen erteilt, eine Art der Qualitätskontrolle ausübt usw.)
e) Der Architekt (der je nach dem Lande mehr oder weniger einflußreich ist, dank seiner Stellung als Vermittler der gesetzlichen Baubewilligung).

Eine Organisation, die auf dem Gebiet der Bauindustrialisierung tätig sein will, muß also entscheiden, wen sie zu ihrem Partner in der Erschließung des Marktes wählen will. In einigen Ländern war die Wahl des Architekten als Ziel dieser Bemühungen sehr erfolgreich. Um Architekten von den Vorteilen der industrialisierten Baumethode zu überzeugen, braucht man viel diplomatischen Takt und handfeste Argu-

regulations, cost limits, lack of materials, etc.). There continues to arrive from abroad a constant flow of interested parties. Obviously the only really worthwhile ware to be sold is the company knowledge in organization and management. Usually, however, a deal is agreed upon whereby the original company sells the technical know-how to the new organization (even if the new organization is fully owned), while the management is provided by local sources.

Only after years of serious difficulties does the original company realise that it has not been selling a worthwhile product (the technique) and, if there is still time, an attempt to help with some management skill is undertaken. It may sound rather strange that industrial organizations are unaware of their most valid product but my own experience shows that this situation is the rule, rather than the exception.

Still very little can be done to define worthwhile strategies: each case has to be analysed on its own terms, and generalisations are not possible.

Tactics are easier to handle. Again not every organization can follow the same approach but a few basic points may be outlined.

Under normal circumstances the ideal situation for any industrialized building organization is to fully control its own market. In a recent report by Masayumi Yokoyama to a Spanish client this fact was stated as follows: "In an industrialized process the building organizations abandon the passive rôle of providers of services for the more active rôle of producers."

Still we must recognise that the path towards controlling a market is not easy and is anyhow obtainable only after a long period of development. Most organizations are therefore obliged to operate within the existing set-up where we identify several participants:

a) The buyer (the person or agency who buys the building)
b) The user (the person or agency who utilizes the building)
c) The builder (the organization which may assemble the building, at times one and the same with the industrialized company)
d) The public authority (which gives building-permits, provides some kind of quality-control, etc.)
e) The architect (who, depending on the country in question, is more or less powerful as the provider of lawful building permits...).

An organization wanting to enter the field of industrialized building has therefore to decide towards which partner he should direct his marketing activities. In a few countries considerable success has been achieved by making the architect the object of the marketing efforts. To convince architects of the advantages of an industrialized building method requires considerable diplomacy and some valid arguments. In the United Kingdom such an approach has been quite successful, while it has failed in the United States.

The user is a complete loss as a reasonable partner. He does not have any decision power and one never knows where to contact him. The buyer of the building may be a public authority, a private organization or an individual. Theoretically he should make the best partner, but often, either through bureaucratic difficulties (as with the public authorities) or through

mente. In Großbritannien war diese Methode sehr wirksam, während sie in den Vereinigten Staaten versagt hat.

Der Benutzer ist für die Organisation als Partner wertlos. Er hat keine Macht, Entscheidungen zu treffen, und man weiß nie, wo man ihn erreichen kann.

Der Käufer des Gebäudes ist entweder eine öffentliche Behörde, eine Privatfirma oder eine Einzelperson. Theoretisch sollte er der beste Partner sein, aber er ist oft schwer zu fassen, entweder wegen bürokratischer Schwierigkeiten, zum Beispiel bei Behörden, oder wegen seiner Unwissenheit, besonders bei Einzelpersonen. Aber auch hier kann viel durch Public Relations oder gezielte Werbung erreicht werden. (Siehe die Siedlungsbauten in Frankreich oder die Schulbauprogramme in Italien, Frankreich und Deutschland.)

Die Hilfe des Bauunternehmers in der Erschließung des Marktes ist nicht immer ein reiner Segen. Natürlich ist er Fachmann auf seinem Gebiet, aber seine Methoden sind in der Regel sehr konservativ. Man kann keinen vollen Erfolg erwarten, aber wie immer gibt es Ausnahmen, wie zum Beispiel das Unternehmen der National Home in den USA.

Ganz gleich nach welchem Vorbild die industrialisierte Bauorganisation sich entwickelt, irgendwann wird ein Bauprogramm entstehen. Auch hier kann es Probleme geben. Die Verkaufsabteilung versucht an jede mögliche Art von Gebäuden heranzukommen – Wohnbauten und Kartoffelspeicher, Kirchen und Schulen, Luxusvillen und Sozialwohnungen. Der Geschäftsleiter oder öfter noch der Leiter der technischen Abteilung bekommt ein fertig entworfenes Projekt oder eine einfache Skizze und soll nun einen Kostenvoranschlag und womöglich einen Entwurf ausarbeiten. Beim konventionellen Bauen ist die Vielfalt der Aufgabe lästig, aber sie läßt sich bewältigen. Für industrialisiertes Bauen müssen ganz deutliche Abgrenzungen festgesetzt werden.

Jede industrialisierte Baumethode ist zur Erreichung der optimalen Leistung auf eine spezifische Art des Gebäudes eingestellt, und obgleich erhebliche Abweichungen technisch möglich sind, ist in der Praxis strikte Disziplin absolut notwendig. Abänderungen kosten Zeit und sind immer teuer. Außerdem ist es in einem komplexen Bausystem oft schwierig, die vollen Implikationen einer Änderung abzuschätzen. Es kann leicht eine Kettenreaktion von Abänderungen geben, die ernsthafte Folgen an Kosten und Zeit nach sich ziehen. Die meisten Organisationen versuchen, ihr Tätigkeitsfeld auf wenige Arten oder nur eine Art von Gebäuden zu beschränken, zum Beispiel Wohnhäuser, Schulanlagen usw. Und doch ist manchmal eine große, außergewöhnliche Aufgabe eine unwiderstehliche Versuchung. Die vorherrschende Notwendigkeit, die Organisation beschäftigt zu halten, läßt oft die festesten Entschlüsse ins Wanken geraten.

Welche Typen von Bauten sind am besten für ein industrialisiertes Bauunternehmen geeignet? Die Erfahrung zeigt, daß alle Gebäude sich für die industrialisierten Baumethoden eignen. Eine deutsche Firma hat sich erfolgreich mit Kirchen befaßt. Hochhäuser sowie Schulanlagen sind Standardprodukte; fast das gleiche gilt für landwirtschaftliche und industrielle Anlagen. Einfamilienhäuser, Ferienbauten und Hotels können ein lohnendes Absatzgebiet sein.

ignorance (as with private individuals) he may be very elusive. Yet again considerable success can be achieved by a well-directed public-relations or advertising campaign. (See the housing-projects in France or the school-building programmes in Italy, France and Germany.)

To utilize the builder as a vehicle for the marketing of the product is a mixed blessing. Clearly the builder knows how to operate in the building field, but at the same time his approach is usually very conservative. Only limited success can be expected, but as usual, exceptions do exist, as is the case with the operation of National Home in the U.S.A.

Whichever pattern the industrialized building organization follows, at one stage or another some kind of building programme will be secured. And here again problems may arise. Naturally the sales department will try to capture any conceivable kind of building from housing to potato-storages, from churches to schools, from luxury villas to public housing.

The manager, or more commonly the head of the technical department, receives a completely-designed project or a simple brief, and is expected to provide a cost-estimate and possibly a design. In conventional building a mixed bag of jobs is a nuisance, but can be dealt with. In industrialized building very clear-cut demarcation lines have to be established. Each industrialized building method is optimized for a specific kind of building and while technically considerable extrapolations may be performed, in practice a rigid discipline is necessary. Changes take time and always cost money. Moreover in a complex building-system it is at times difficult to appreciate the full implications of a modification. A chain-reaction of changes may well be started with serious consequences in cost and time. Most organizations try to limit their field of activity to a few or to one kind of building (housing, educational facilities, etc.). Still a big, if exotic, job is at times an irresistable temptation. The imperious need of keeping the organization busy can weaken the most stringent resolutions.

Which kind of building is best suited for an industrialized building venture? Experience has shown that practically all buildings can be successfully erected with an industrialized approach. A German company has successfully dealt with churches. Multi-storey housing is a standard staple, the same as educational facilities. Agricultural and industrial facilities are a common product; one-family housing, holiday-housing and hotels can offer a worthwhile market.

The operation method alters, however, according to the kind of building required. In dealing with public buildings such as schools, a limited amount of contacts can be expected but a very high ratio of signed contracts is common (about 60%). Also the value of the single building can be relatively small and a starting delay or a cancellation may not affect too seriously the production programme. The situation changes with the multi-storey housing projects or one-family housing developments (300–500 dwellings per job). Here just a few contacts take place, and very few contracts are signed, but the dropping-out of one project may seriously upset the whole activity of the organization. Different again is the case of the individual one-family houses (or agricultural and industrial facilities): a huge number of enquiries and

Die Vertriebsmethode hingegen ändert sich je nach der Art des Bauprojektes. Bei öffentlichen Gebäuden, wie zum Beispiel Schulen, kann eine begrenzte Anzahl von Kontakten erwartet werden, aber eine sehr hohe Rate unterschriebener Verträge ist die Regel (ungefähr 60%). Auch ist der Wert des einzelnen Gebäudes oft ziemlich gering, und ein anfänglicher Aufschub oder eine Annullierung brauchen das Produktionsprogramm nicht unbedingt ernsthaft zu beeinträchtigen. Die Situation ist anders bei Hochhäusern oder Einfamilienhaussiedlungen (300–500 Wohnungen pro Auftrag). Hier werden nur wenige Anfragen stattfinden, und sehr wenige Verträge werden unterschrieben, aber das Ausfallen nur eines Projektes kann die gesamte Tätigkeit der Organisation beeinträchtigen. Wieder anders ist es bei Einfamilienhäusern oder auch bei landwirtschaftlichen und industriellen Anlagen: die Regel ist eine Riesenanzahl von Anfragen und Kontakten mit einer sehr niedrigen Rate erfolgreich abgeschlossener Projekte (ungefähr eins zu zehn). Jedoch hat der geringe Wert jedes Einzelobjekts den Vorteil, daß der Verlust oder Aufschub einiger Bauaufträge weniger spürbar wird.

Je nach Art der Gebäude, um die es sich handelt, können verschiedene Verfahren angemessen sein. Für einen weitauseinanderliegenden Markt von Einfamilienhäusern oder kleineren Industrieanlagen wird die Zusammenarbeit mit einheimischen Bauunternehmern unvermeidlich. Es würde enorme Versorgungs- und Nachschubprobleme schaffen, jedes Einzelobjekt mit Montagekolonnen zu versehen. Allerdings ist der Organisation, die sich große Projekte zur Aufgabe macht, oft besser mit ihren eigenen Baumannschaften gedient. Die Zusammenarbeit mit anderen Bauunternehmern hat für die Organisation den Vorteil einer leichten Dauerbelastung, aber die Nachteile von erhöhten Kosten und Mangel an Kontrolle. Eigene Montagemannschaften tun einen besseren Dienst, aber sie sind eine große Belastung für die Firma in flauen Zeiten. In der Praxis versuchen die meisten Organisationen, je nach Bedarf beide Methoden anzuwenden.

Was die Produktion betrifft, ist es vernünftig, den größten Teil auf leicht verkäufliche Standardprodukte mit niedrigem Gewinn zu konzentrieren und ein einträglicheres Spezialprodukt als die Hauptprofitquelle zu haben. Kombinationen dieser Art sind Sozialwohnungen und Krankenhäuser, Schulen und Bürohäuser, Einfamilienhäuser und Siedlungen usw.

Auch auf dem Gebiet der Montage ist eine Mischung beider Methoden üblich. Die Organisation hat eine beschränkte Anzahl von Montagemannschaften, und nötigenfalls arbeitet sie mit anderen Bauunternehmern zusammen.

Eine besondere Situation ist das Marketing von «Know-how»-Lizenzen an ausländische Unternehmer. Gewöhnlich liefert der Verkäufer, außer Personal und gelegentlich einigen speziellen Bauwerkzeugen, technisches «Know-how» gegen eine feste Anzahlung, manchmal in Form von Geschäftsanteilen und Lizenzhonoraren. Nach meiner Erfahrung lohnt sich diese Abmachung nur für den Verkäufer. Die technische Lösung muß immer den Ortsverhältnissen angepaßt werden, und nachdem die Abänderungen ausgeführt sind, bleibt wenig von dem ursprünglichen «Know-how» übrig. Praktisch alle existierenden Fälle dieser Art bestätigen die Regel, daß eine einheimische Entwicklung billiger und zufriedenstellender ist und daß Lizenzverträge auf hochspezialisierte Herstellungsmethoden beschränkt und auch dann nur mit äußerster Vorsicht abgeschlossen werden sollten.

contacts is common with a very low rate of successfully-completed projects (about 1 in 10). However, the small value of each individual project offers the advantage of reducing the impact of some lost or delayed jobs.

According to the kind of building the organization is concerned with, different operational forms may be suitable. For a widely-dispersed market of one-family houses or small industrial facilities a collaboration with local builders becomes inevitable. To attempt to provide erection teams for each project would cause enormous logistical problems. Viceversa when big projects are the aim, the organization's own erection teams may prove to be more satisfactory. The collaboration with outside builders offers the advantage of a light permanent-structure load, but the disadvantages of heavier costs and lack of control. Own assembly crews are more effective but easily become a mill-stone around the neck of the company during poor construction periods.

In practice most successful organizations try to operate on a reasonable mixed basis.

A satisfactory pattern is that of marketing a standard, easy-to-sell product with little financial return to provide the bulk of the production, and a more rewarding, specialized product as the major source of profit. (Combinations of this kind are public-housing and hospitals, schools and office-building, one-family housing and public-developments, etc.)

On the assembly side, a mixed approach is common. The organization owns a limited number of assembly crews and collaborates with outside builders whenever necessary.

A special situation is that of marketing a know-how license to a foreign operator. Usually the selling-company provides technical know-how, plus staff and possibly some special equipment, in exchange for a fixed down-payment (sometimes in the form of stocks) and a royalty. In my experience this arrangement is only worthwhile for the selling company. The technical solution must always be modified to suit local conditions. And by the time the modifications are carried through little remains of the original know-how. From practically all existing cases of this kind the rule can be established that a local development is cheaper and more satisfactory, and that licensing should be limited to highly specialized production methods, and even then with the utmost caution.

4.5 The subsidiary industry

A major development often overlooked in the industrialization of building during the last years has been the success of manufacturers producing semi-finished products or sub-systems.

In actual fact this development may in the long run be of more relevance to the industrialization of building than the local successes of this or that modular or prefabricated system. It is already today possible for a willing architect to design, and for a contractor to build, a complete building out of components or sub-systems which are obtainable through a manufacturer's catalogue.

Unfortunately such components or sub-systems are not dimensionally interrelated even when the modular discipline has been applied. Complete industrialized structures are available (in concrete or in steel) both in the U.S.A. and in Europe. Floor sub-systems can be chosen from catalogues (Durisol, Siporex, Robertson, Acieroid, Davun and Silicalcite, only to mention a few). Curtain-wall or window sub-systems are available by the dozen, as are all kinds of partition sub-systems (FEAL, Hausermann, Fillod, Bellrock, etc.). In addition to the above it has become possible to buy complete sanitary blocks, including fixed and semi-fixed appliances made out of classic materials like ceramic or out of the most modern plastic.

But even more relevant is in my opinion the existence of companies who solve the classic problem of the industrialized building: the joint. In the United States special "fasteners" industries can provide the most complex kind of mechanical joint capable of dominating the most extravagant tolerances. In Europe all possible kinds of mastic are available (backed – if required – by bank guarantees) to ensure the impermeability of a building.

All these young enterprises (possibly none of them is older than twenty years) are slowly growing to form the backbone of the industrialization of building and their quiet operational form is out-flanking the resistance of the most conservative professionals and authorities.

4.6 Possible organizational forms

In an overall scheme for the possible organization of the industrialization of building I think two sections should be accepted as nearly certain, while the intermediate area may well be left for many years to come, in a state of open competition.

At the top of the organizational scheme we need a strong research and development-activity fully devoted to the problem of the creation of a satisfactory human urban environment.

On the other hand we can already see the development of a range of manufacturers of equipment (from cranes to nailing machines) of components (windows, sanitary and mechanical equipment, panels, etc.) and of sub-systems (curtain walls, structures, etc.).

In between, all kinds of different organizations may operate, from the prefabricator of buildings to the organizing service, from the rationalized builder to the

Zwischen diesen beiden Positionen können alle möglichen organisatorischen Vorgehen gewählt werden, von der Herstellung von Fertigbauten bis zum Organisierungsdienst, vom rationalisierten bis zum ausrüstungsintensiven Bauunternehmen. Man kann heute noch nicht mit Bestimmtheit sagen, welche organisatorische und technologische Form als die bestgeeignete hervorgehen wird. Nur eines steht fest: allen Methoden der Bauindustrialisierung muß die Chance gegeben werden, sich zu bewähren.

Zu oft in den letzten zwanzig Jahren haben Regierungs- und Privatstellen beschlossen, blindlings einer bestimmten Art der Bauindustrialisierung zu folgen (siehe das russische Wohnungsprogramm oder das britische Schulbauprogramm), und haben dabei vielleicht interessante Entwicklungen aufgehalten und eine führende Rolle preisgegeben.

Zweifellos ist die Hauptschwäche des heutigen Standes der Bauindustrialisierung der Mangel an ernster Forschungstätigkeit. Für einen Politiker ist die Versuchung zu groß, den Weg des geringsten Widerstandes zu wählen und die Schaffung von einigen Fabrikationsbetrieben zu befürworten. Auf dem Gebiete des Bauens wie auf anderen Gebieten ist die Technologie ein zu gefügiges Werkzeug geworden, das mangels eines geplanten Anwendungsprogrammes verschiedene Manipulationen ermöglicht.

equipment-oriented constructor. It is too early to identify with certainty which organizational and technological form will prove to be the most satisfactory. In fact the only safe recommendation we can give today is to give all forms of industrialized building a chance to come to grips with reality.

Too often in the past twenty years, governments and private authorities have decided to follow blindly a specific form of industrialization (see the Russian effort in housing or the British effort in school-buildings) in this way hindering possible interesting developments and loosing in the process a leading position.

Certainly the major weakness of the industrialization effort lies in the absence of a serious research-effort. Today it is too tempting for a politician to take the short cut of sponsoring the creation of a few manufacturing units, and... to hell with the consequences! In building as in other fields, technology has become a too readily pliable tool, allowing all kinds of manipulations in the absence of a planned programme of application.

5. Konkrete Beispiele

5. Case Studies

5.1 Einleitung

Zur Illustration der gegenwärtigen Situation in der Entwicklung der Bauindustrialisierung wählte ich einige Firmen aus und versuchte, die Besonderheiten ihres Betriebes zu analysieren.
Die Auswahl der Firmen erfolgte nach folgenden Kriterien:
a) Die Firmen sollten ein durchschnittliches Bild ihrer Tätigkeit im eigenen Land geben.
b) Die Firmen sollten wirtschaftlich erfolgreich sein und eine ausreichend lange Tätigkeit aufweisen können.
Bei den meisten Ländern war die Wahl gar nicht leicht. Die ausgewählten Firmen stellen aber meiner Meinung nach einen aufschlußreichen Querschnitt durch die verschiedenen Arbeitsweisen dar und erweisen sich als ausreichend repräsentativ.
Die Analyse gliedert sich bei jeder Firma in sechs Hauptthemen:
a) Name und Beschreibung der Firma
b) Beschreibung des Produktes
c) Produktion
d) Personal
e) Arbeitsweise
f) Allgemeine Informationen
Die gesammelten Informationen kamen durch persönliche Besuche und in zwei Fällen durch briefliche Kontakte zustande. In allen Fällen hat aber immer die Firma selbst die Informationen geliefert; ich habe meinerseits nicht versucht, diese Auskünfte zu modifizieren.
Soweit ich also von außen urteilen kann, ist die Information korrekt, auch wenn die Möglichkeit von Fehlern nicht ausgeschaltet werden kann.
Die Auswahl des Bildmaterials überließ ich vollkommen der Diskretion der Firma. Wir können deshalb annehmen, daß es das Bild des Produktes widerspiegelt, das die Firma nach außen vertreten will.
Jede Firma wird in einer graphischen Darstellung zusammengefaßt, damit sich der Leser eine konkrete Vorstellung der Arbeitsweise der betreffenden Firma machen kann.
In der Darstellung auf Seite 68 sind die Tätigkeiten, womit die Firma sich direkt befaßt, in Schwarz, diejenigen, die sie mit anderen Firmen teilt, schraffiert dargestellt (dieses Diagramm faßt also die gesammelte Information und deren Einschätzung durch den Verfasser zusammen).

5.1 Introduction

In order to exemplify the present situation in the development of the industrialization of building I have chosen a few organizations and attempted to analyse their way of operating. The criteria for the selection have been:
a) The organizations should represent an average picture of their activity in their own country.
b) The organizations should be commercially successful and in operation for a sufficient period of time.
Obviously for most countries the choice was not very easy. However I think that the selected companies offer a good picture of different operational forms and are sufficiently representative.
For each organization the analysis is divided into six main subjects:
a) Name and description of the company
b) The description of the product
c) Production
d) Staff
e) Operational forms
f) General information.
The information I have collected has been obtained either through personal visits and in two cases through mail contact. In all cases the information has been provided by the organization itself, without any attempt at modification on my part.
Still, as far as I can judge the situation from outside, the information given is correct, even if some material mistakes cannot be excluded.
The choice of the graphic material has been left entirely to the discretion of the organization. We can therefore assume that the photographic material reflects the image of the product the organization wishes to present to the outside world.
For each organization a graph attempts to give a visual summary of the operational form of the organization.
In the graph on page 68 the activities in which the organization is directly involved are shown in black and hatched the ones in which it is in collaboration with others (this graph is the result of the information given and my own assessment).

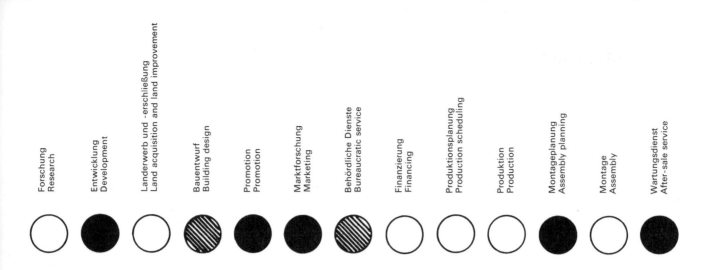

Die Firma ist direkt verantwortlich für: Entwicklung, Promotion, Marketing, Montageplanung, Wartungsdienst.
Die Firma ist gelegentlich verantwortlich für: Bauentwurf, bürokratische Dienstleistungen.

Einige Begriffe in der graphischen Darstellung müssen näher erklärt werden, um Mißverständnissen vorzubeugen.
Der Begriff «Forschung» bezeichnet hier lediglich die Grundlagenforschung technologischer oder anderweitiger Art. Was einige Firmen als Forschung bezeichnen, ist eigentlich bereits schon Entwicklung oder bestenfalls angewandte Forschung. Die Zusammenstellung der Diagramme beruht also auf dem enger gefaßten Sinn des Wortes.
Der Begriff «Promotion» bezeichnet den Vorgang der Entwicklung und Förderung neuer Siedlungen oder ähnlicher Unternehmungen. Dieser Begriff ist äußerst verwirrend (ein Synonym wäre das Wort «Entwicklung»). Das beste Beispiel für einen «Promoter»: Er kann durch seine Verbindungen oder durch persönliches Eigentum einen Grundstücksbesitzer überzeugen, ein Stück Land (etwa ein Gebiet in der Nähe von Paris oder eine Insel im Karibischen Meer) für den Bau einer luxuriösen Wohnsiedlung oder eines neuen Touristenzentrums zur Verfügung zu stellen. Nach dem Erwerb des Grundstücks (einem geschickten Promoter gelingt der Landerwerb, ohne einen Rappen auszugeben) macht er sich nun an die Aufgabe, einen Finanzmann von der Seriosität des Projektes zu überzeugen. Besitzt der Promoter noch Optionsrechte auf das Grundstück und ein ansprechendes (und flexibles) architektonisches Projekt, wird sich dieser zweite Teil des ganzen Prozesses ohne besondere Schwierigkeiten abwickeln. Schließlich kommt noch der Unternehmer zum Zug. Sieht er, daß eine Bank hinter dem Projekt steht, kennt er den Ruf des Promoters und ist er von der Zuverlässigkeit des Projektes überzeugt, dann stellt er seine Dienste zur Verfügung.
Wenn alles am Schnürchen läuft, werden nun die Wohnungen, Häuser, Verkaufszentren, oder was immer es sei, rechtzeitig verkauft, um die Fabrikanten zu bezahlen, das heißt, um sich an den Finanzierungsplan halten zu können. Der Promoter, der Grundstücksbesitzer, der Unternehmer und die Bankgesellschaft haben dann ihr Ziel erreicht. Es können aber auch Schwierigkeiten auftauchen, Fabrikanten werden ge-

Organization involved in: development, promotion, marketing, assembly planning, after-sale service.
Organization occasionally involved in: building design, bureaucratic service.

Some of the terms used in this graph need explaining to avoid misunderstandings.
Under the term "research" I include only basic research of technological or other nature. What some companies consider research is in fact development, or, at best, applied research. It is with the restrictive meaning of the term that I have compiled the graphs. Promotion means the activity of developing and inspiring new settlements or ventures of all kinds. A lot of confusion exists around this term (an alternative is the word "development"). In its purest form a "promoter" can be visualized as the gentleman who, through his connections, or because of his personal fortune, convince a certain land-owner to make available some land (an area near Paris, an island in the Carribean) on which to build a luxury residential settlement, or a new tourist complex. Having obtained the land (if the promoter is very efficient, he will obtain it without having to invest a penny) he now proceeds to convince a financer of the soundness of the project. Being in possession of some kind of option on the land and of an appealing (and flexible) architectural project this second part of the exercise may not prove too difficult. Finally a contractor is brought into the game who, seeing the project is backed by a banker, knowing the reputation of the "promoter" and convinced again of the soundness of the project undertakes to provide his services. If everything goes well the flats, the cottages, the shopping centre or what maybe will be sold in accordance with the payments due to the suppliers, or within the financing schedule. The promoter, the land owner, the contractor and the banker will then be richer people. Otherwise troubles may arise, suppliers may become nasty, clients may turn to the courts...
Obviously the above example is not typical of all promoters, but even when the "promoter" becomes a prestigious, overstaffed company, the rules are basically the same: a good promoter creates buildings or better "business" with other people's money.
It must be well understood that "promotion" is a very

hässig, und Kunden wenden sich an die Gerichte...
Dieses Beispiel gilt selbstverständlich nicht für alle Promoters. Aber auch wenn der «Promoter» eine berühmte, mit Angestellten überdotierte Firma ist, bleibt sich dieser Vorgang grundsätzlich gleich: Ein guter Promoter schafft Gebäude – besser gesagt, er macht «Geschäfte» mit anderer Leute Geld.

Man muß verstehen, daß die «Promotion» einen vollkommen legitimen und notwendigen Vorgang innerhalb der Bauindustrie darstellt und daß die finanziellen Schwierigkeiten, in welche die Promoters gelegentlich geraten, meistens auf eine unzulängliche Kalkulation zurückzuführen sind.

Der Begriff «Finanzierung» bezieht sich auf jene Tätigkeit, durch die für den Kauf des Produktes Geld aufgetrieben wird, und nicht auf die einfache Finanzierung, die jedem Geschäftsprozeß zugrunde liegt und durch die zeitliche Lücke zwischen der Fertigstellung einer Dienstleistung und der Zahlung bedingt ist. Anders gesagt geht es also dann um die «Finanzierung», wenn diese Tätigkeit eine zusätzliche Profitquelle darstellt. Alle Firmen finanzieren ihre Kunden in einem gewissen Sinn; aber dann werden die Kosten der Finanzierung einfach zu den Kosten des Produktes hinzugeschlagen. Wenn eine Firma eine Schule für die Regierung baut und weiß, daß Zahlungen nach zwölf Monaten erfolgen, wird diese Firma in irgendeiner Form Kredit von einer Bank verlangen müssen und die zusätzlichen Kosten dieses Geldes zum Verkaufspreis des Produktes hinzuschlagen.

Eine solche Firma befaßt sich nicht mit «Finanzierung»; sie nimmt einfach einen ziemlich unerwünschten, aber unvermeidlichen Umstand in Kauf. Ganz anders sieht es bei einer Firma aus, die nach dem Bau eines Wohnblocks die Wohnungen verkauft und den Käufern langfristige Kredite gewährt. In diesem Fall kann der Profit des gesamten Unternehmens die Finanzierungstätigkeit sein.

Mit dem Begriff «bürokratische Dienstleistungen» möchte ich all jene Tätigkeiten zusammenfassen, mit denen sich eine Firma befassen muß, um Bauerlaubnisse zu bekommen, um den diversen Anforderungen einer ganzen Reihe von Vorschriften zu genügen, um das Gebäude mit elektrischen, Gas-, Straßen- und Telephonverbindungen usw. zu versorgen.

Dieser Bereich ist in sich selber undankbar; ein potentieller Kunde aber, der sich nicht in den erdrückenden Subtilitäten unserer Bürokratien verstricken will, könnte sich für ein bestimmtes Unternehmen entscheiden, gerade weil es diese Angelegenheiten auch erledigt.

Die «bürokratischen Dienstleistungen» bilden ein geschätztes Verkaufsargument.

legitimate and necessary activity within the building industry, and that some of the financial "cracks" in which promoters are occasionally involved are due mostly to poor calculations.

With the term "financing" I wish to define the activity through which money is provided for the buying of the product, and not the simple financing inherent to every business and caused by the time-gap between the provision of a service and the payment for the same. In other words, I consider that an organization is involved in "financing" when this activity is an additional source of profit. All organizations are financing their customers to some extent, but the cost of the financing is simply added to the cost of the product. When an organization builds a school for a government knowing that payments will be received after twelve months, this organization will ask some form of credit from a bank and will add the cost of this money to the sale-price of the product. This organization is not involved in "financing", it simply accepts a rather undesirable but necessary circumstance. Very different is the case of an organization which, after having produced an apartment-house will sell the apartments and will provide long-range credit to the buyers. In this case the profit of the operation may well be within the financing activity.

Under the term "bureaucratic service" I wish to cover all the activities an organization may have to indulge in, in order to obtain building permits, to satisfy the requirements of all possible kinds of inspectors, to provide the building with electrical, gas, road, telephone connections, etc.

Obviously this kind of activity is in itself unrewarding, but a potential client lost in the nightmarish niceties of our bureaucracies may decide in favour of a particular company simply because it provides this kind of help.

The "bureaucratic service" becomes a helpful sales argument.

National Home Corporation

Lafayette, Indiana, Earl Avenue, 47902 USA

Tätig seit: 1940
Tätig in: Vereinigte Staaten
Art des Produktes: Häuser und Wohnungen, «Mobile Homes», Schulhäuser und gewerbliche Bauten, wobei Häuser und Wohnungen ungefähr 60%, «Mobile Homes» 30% der Gesamtproduktion ausmachen

Beschreibung des Produktes

Fabrikhergestellte Bauten, basierend auf konventioneller Holztechnik, mit einer allgemein ansprechenden Gestaltung, die großzügige Alternativen in der Anordnung bietet. Nur ein- oder zweigeschossig

Produktion 1970

Montierte Bauten pro Jahr: 28 000 Einheiten
Gesamtwert der Verkäufe pro Jahr: 85 000 000 Dollar
Gesamtkosten für Arbeitskräfte pro Jahr: 37 171 000 Dollar
Gesamtwert des investierten Kapitals bis heute: 37 500 000 Dollar

Personal

Leitung: 483 Personen
Entwerfer und Techniker: 94 Personen
Technisches Personal für die Produktion: 86 Personen
Fabrikarbeiter und Montagemannschaft auf der Baustelle: 2908 Personen (davon 826 gelernt, 1149 angelernt und 933 ungelernt)
Büropersonal: 687 Personen
Transport und Versand: 250 Personen
Sonstiges: –
Gesamtbestand des Personals: 4500 Personen

Arbeitsweise

Finanzierung: Je nach Wunsch des Bauunternehmers, obwohl 70% der Finanzierung durch National Home erfolgte. National Home hat eine eigene Finanzierungsgesellschaft und kann jeden konzessionierten Unternehmer mit Finanzkapital versehen
Bautätigkeit: National Home arbeitet mit 1250 konzessionierten Unternehmern, die sich auf 38 Staaten verteilen. Die Firma besitzt eine eigene Baugesellschaft, tätig an 30 Orten, errichtet annähernd 20% aller Bauten
Entwurf der Komponenten: Eigener Entwurf
Entwurf des Baus: National Home gibt gelegentlich Aufträge an unabhängige Architekten, um ein neues Wohnbaumodell zu entwerfen. Daraufhin entwickelt das interne technische Personal den ursprünglichen Entwurf
Produktion der Komponenten: In den firmaeigenen Werkhallen
Marketing: Durch konzessionierte Unternehmer oder die firmaeigene National Home Construction Corporation. National Home verkauft nur an Unternehmer
Kauf und Erschließung des Grundstückes: National Home kauft und erschließt manchmal auch Land für zukünftige Bauprojekte

Allgemeine Information

Durchschnitt pro verkauftes Haus: 1200 Quadratfuß
Durchschnittlicher Verkaufspreis pro Haus: 18 000–19 000 Dollar
Durchschnittlicher Verkaufspreis pro «Mobile Home» (an Händler): 5000 Dollar
Wert des fabrikhergestellten Baus: Für Häuser, die innerhalb von 5 Tagen montiert werden: 75% des Schlußwertes. Für Häuser, die innerhalb von drei Wochen montiert werden: 25% des Schlußwertes

National Home Corporation

Lafayette, Indiana, Earl Avenue, 47902 U.S.A.

Organization in operation since: 1940
Organization operating in: United States
Line of product: Houses and apartments, mobile homes, schools and commercial buildings, whereby houses and apartments represent approximately the 60% of the total and mobile homes the 30%

Description of product

A factory-manufactured building, based on a conventional wood technology, making use of a mass-appealing design offering substantial layout alternatives. One or two storeys only

Production 1970

Assembled buildings per year: 28 000 units
Total value of sales per year: 85 000 000 dollars
Total cost of manpower per year: 37 171 000 dollars
Total value of capital investment up to date: 37 500 000 dollars

Staff

Management: 483 persons
Designers and technical staff: 94 persons
Production technical staff: 86 persons
Factory workers and site assembly crews: 2908 persons (of which 826 skilled, 1149 semiskilled and 933 unskilled)
Clerical staff: 687 persons
Transport and shipment staff: 250 persons
Others: –
Total staff employed: 4500 persons

Way of operation

Financing: Optional with the builder, however 70% of the financing was provided by National Home. National Home owns its own financing company and can provide any franchised builder with financing capital
Building activity: National Home operates through 1250 franchised builders in 38 States. The organization owns its own Construction Corporation, who operates in 30 locations and erects approximately 20% of all erected buildings
Design of the components: Own design
Design of the building: National Home hires occasionally an outside architect to design a new housing model. Subsequently the internal technical staff develops the original design
Production of the components: National Home own factories
Marketing: Through the franchised builders or the fully owned National Home Construction Corporation. National Home sells only to builders
Purchase and development of land: Occasionally National Home purchases and improves land for future development

General information

Average square footage of sold house: 1200 sq.ft
Average selling price of house: 18 000–19 000 dollars
Average selling price of mobile home (to dealer): 5000 dollars
Value of the building produced in the factory: For houses to be erected within 5 days: 75% of final value. For houses to be erected within three weeks: 25% of final value

Forschung Research	Entwicklung Development	Landerwerb und -erschließung Land acquisition and land improvement	Bauentwurf Building design	Promotion Promotion	Marktforschung Marketing	Behördliche Dienste Bureaucratic service	Finanzierung Financing	Produktionsplanung Production scheduling	Produktion Production	Montageplanung Assembly planning	Montage Assembly	Wartungsdienst After-sale service

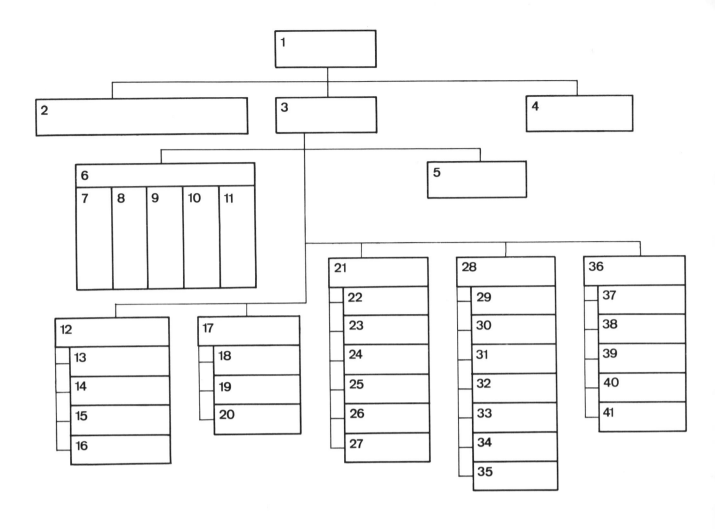

1 Verwaltungsrat	1 Board of directors
2 Ständiger Ausschuß: Interessenkonflikte – Kompensation – Prämienabschlüsse	2 Standing committees: Conflict of interest – compensation – stock options
3 Präsident des Verwaltungsrates	3 Board chairman
4 Exekutivausschuß	4 Executive committee
5 Leitungsausschuß	5 Management committee
6 Personal	6 Staff
7 Vizepräsident Corporate Marketing	7 Vice-president corporate marketing
8 Programm-Manager «Operation Breakthrough»	8 Programme manager «Operation Breakthrough»
9 Direktor für Produktentwicklung	9 Director product development
10 Direktor für Forschung und Technologie	10 Director technology and research
11 Direktor der Personalverwaltung	11 Director of staff administration
12 Exekutiver Vizepräsident: Herstellung	12 Executive vice-president: Manufacturing
13 Vizepräsident: Horseheads Division	13 Vice-president: Horseheads Division
14 Präsident: National Home Corporation of Georgia	14 President: National Home Corporation of Georgia
15 Präsident: National Home Corporation of Virginia	15 President: National Home Corporation of Virginia
16 Vizepräsident: Tyler Division	16 Vice-president: Tyler Division
17 Präsident: National Home Acceptance Corporation	17 President: National Home Acceptance Corporation
18 Mayflower Lebensversicherung	18 Mayflower Life Insurance
19 National Home Realty Corporation	19 National Home Realty Corporation
20 Präsident: Southwest Title Insurance Company	20 President: Southwest Title Insurance Company
21 Präsident: National Home	21 President: National Home
22 Vizepräsident: Lafayette Division und Town House Parks	22 Vice-president: Lafayette Division and Town House Parks
23 Präsident: National Home Corporation Structures	23 President: National Home Corporation Structures
24 Präsident: W. G. Best Homes	24 President: W. G. Best Homes
25 Präsident: National Mobile Homes	25 President: National Mobile Homes
26 Präsident: National Fiber Glass	26 President: National Fiber Glass
27 Manager: New Albany Cabinet Plant	27 Manager: New Albany Cabinet Plant
28 Exekutiver Vizepräsident	28 Executive vice-president
29 Präsident: National Home Construction Corporation	29 President: National Home Construction Corporation
30 Exekutiver Vizepräsident: Firmeneigene Abteilung für Einfamilienhäuser	30 Executive vice-president: Company-owned Single-Family Home Division
31 Exekutiver Vizepräsident: Abteilung für spezielle Projekte	31 Executive vice-president: Special Projects Division
32 Exekutiver Vizepräsident: Abteilung Subvention und Spekulationsgeschäfte	32 Executive vice-president: Subsidy and Venture Division
33 Exekutiver Vizepräsident: Marktforschung	33 Executive vice-president: Marketing
34 Präsident: Burton Division	34 President: Burton Division
35 Exekutiver Vizepräsident: Grundstücke	35 Executive vice-president: Land Division
36 Exekutiver Vizepräsident: Finanzen	36 Executive vice-president: Finance
37 National Resort Communities	37 National Resort Communities
38 Vizepräsident und «Controller»	38 Vice-president and controller
39 Vizepräsident für Entwicklung der Gesellschaft	39 Vice-president corporate development
40 Sekretariat – Schatzmeister	40 Secretary – treasurer
41 Interne Rechnungsprüfung	41 Internal audit

Delta Homes Corporation

Elkhart, P.O. Box 606, Indiana 46514, USA

Tätig seit: 1966
Tätig in: Vereinigte Staaten
Art des Produktes: «Mobile Homes»

Beschreibung des Produktes

Ein fabrikhergestelltes Haus auf Stahlgerüst, das mit Rädern versehen ist. Manchmal werden zwei fabrikhergestellte Einheiten örtlich zusammenmontiert, um eine doppelt breite Einheit zu erhalten. Holzrahmensystem, das der konventionellen amerikanischen «Balloon Frame»-Technik folgt

Produktion 1971

Montierte Gebäude pro Jahr: 107 500 Quadratfuß
Verkaufspreis des fertigen Hauses pro Quadratfuß (1): 9.50 Dollar
Gesamtumsatz pro Jahr: 11 375 000 Dollar
Gesamtunkosten für Arbeitskräfte pro Jahr: 1 700 000 Dollar
Gesamtwert des investierten Kapitals (2): 293 000 Dollar

Personal

Leitung: 6 Personen
Entwerfer und Techniker: 4 Personen
Technisches Personal für die Produktion: 2 Personen
Fabrikarbeiter: 90 Personen (Saisondurchschnitt)
Büropersonal: 6 Personen
Transport und Versand (3): 6 Personen
Verkauf: 5 Personen
Montagearbeiter auf der Baustelle: 3 Personen
Andere: –
Gesamtbestand des Personals: 150 Personen

Arbeitsweise

Finanzierung: Delta Homes finanziert nicht
Marketing: Delta Homes verkauft an Händler (im ganzen 106)
Tätigkeit auf der Baustelle: Durch Händler oder durch Entwickler der «Mobile Home»-Siedlung
Entwurf der Komponenten: Eigener Beitrag begrenzt, es werden hauptsächlich allgemein erhältliche standardisierte Elemente gekauft
Entwurf des Gebäudes: Eigener Entwurf
Werbung: Durch die Firma und die Händler

Allgemeine Information

Maximale Produktionsleistung: 1750 Einheiten pro Jahr oder annähernd 10 Einheiten pro Tag
Durchschnittliche Größe der Einheit: 65×12 Fuß
Maximale Größe der Einheit: 70×14 Fuß
Durchschnittlicher Verkaufspreis einer Einheit: 7000 Dollar
Durchschnittlicher Verkaufspreis einer doppelten Einheit: 10 000 Dollar
Inneneinrichtung: Alle Einheiten werden vollständig möbliert und eingerichtet verkauft

Bemerkungen

(1) Für eine standardisierte Einheit, Händlerpreis; Land und Wasser, Elektrizitätsanschlüsse nicht inbegriffen
(2) Die Firma arbeitet in gemieteten Gebäuden
(3) Gelegentlich werden Fahrzeuge gemietet

Delta Homes Corporation

Elkhart, P.O. Box 606, Indiana 46514, U.S.A.

Organization in operation since: 1966
Organization operating in: United States
Line of product: Mobile homes

Description of product

A factory-produced house on a steel frame provided with wheels. Occasionally two factory-produced units are assembled together on site to provide a double wide unit. Wooden structure following the conventional American balloon frame technic

Production 1971

Assembled square footage of building per year: 107 500 sq.ft
Sales price of sq.ft of finished building (1): 9.50 dollars
Total value of sales per year: 11 375 000 dollars
Total cost of manpower per year: 1 700 000 dollars
Total value of capital investment (2): 293 000 dollars

Staff

Management: 6 persons
Designers and technical staff: 4 persons
Production technical staff: 2 persons
Factory workers: 90 persons (seasonal average)
Clerical staff: 6 persons
Transport and shipment staff (3): 6 persons
Sales force: 5 persons
Site assembly crews: 3 persons
Others: –
Total staff employed: 150 persons

Way of operation

Financing: Delta Homes does not provide any financing
Marketing: Delta Homes sells to dealers (the company deals with 106 dealers)
Site activity: Provided by the dealers or by the mobile-home park developers
Design of the components: Limited own design, most of the hardware is bought from standard available products
Design of the building: Own design
Advertising: By the company and through the dealers

General information

Maximum production capability: 1750 units per year or approximately 10 units per day
Average size of unit: 65×12 ft
Maximum size unit: 70×14 ft
Average selling cost of a unit: 7000 dollars
Average selling cost of a double unit: 10 000 dollars
Furnishing: All units are sold fully furnished and equipped

Remarks

(1) For a standard unit, price to the dealer, exclusive of land and water, electricity, connections
(2) The company operates in leased premises
(3) Occasionally outside tractors are hired

79

Forschung / Research	Entwicklung / Development	Landerwerb und -erschließung / Land acquisition and land improvement	Bauentwurf / Building design	Promotion / Promotion	Marktforschung / Marketing	Behördliche Dienste / Bureaucratic service	Finanzierung / Financing	Produktionsplanung / Production scheduling	Produktion / Production	Montageplanung / Assembly planning	Montage / Assembly	Wartungsdienst / After-sale service
○	●	○	●	○	●	○	○	●	●	○	○	○

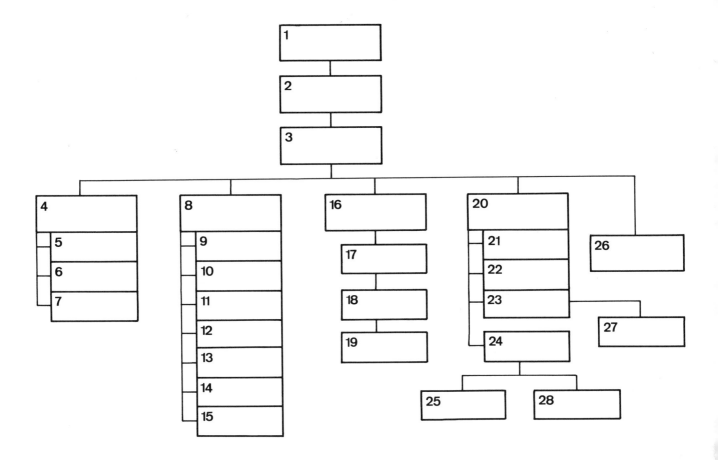

1	Verwaltungsrat	Board of directors
2	Präsident	President
3	Exekutiver Vizepräsident	Executive vice-president
4	Vizepräsident: Finanzen	Vice-president: Finances
5	Buchhaltungsleiter	Senior accountant
6	Fakturierung	Billing
7	Personal	Personnel
8	Vizepräsident: Verkauf	Vice-president: Sales
9	Transportleiter	Transport manager
10	Verkaufsleiter: Mideastern	Sales manager: Mideastern
11	Verkaufsleiter: Midwestern	Sales manager: Midwestern
12	Verkaufsingenieur	Sales engineer
13	Verkaufsingenieur	Sales engineer
14	Verkaufsinspektoren	Sales inspectors
15	Verkaufsingenieur der Abteilung für Modularen Wohnungsbau	Sales engineer Modular Housing Division
16	Vizepräsident: Herstellung	Vice-president: Manufacturing
17	Vizebetriebsvorsteher und Inspektor	Assistant plant superintendant and inspector
18	Gruppenführer	Group leaders
19	Eigenes Personal	Direct labour
20	Vizepräsident: Produktentwicklung	Vice-president: Product development
21	Produktentwicklung und Kosten	Product development and cost
22	Einkauf	Purchasing
23	Entwurf, Konstruktion, Technisches Zeichnen	Design, engineering, drafting
24	Zentrale Annahme	Central receiving
25	Unterhalt	Maintenance
26	Abteilung für Modularen Wohnungsbau, Generaldirektor	Modular Housing Division, general manager
27	Technisches Zeichnen	Drafting
28	Freie Mitarbeiter	Indirect labour

IPI S.p.A.
Mailand, Italien

IPI S.p.A.
Milan, Italy

Tätig seit: 1942
Tätig in: Italien, Griechenland, Deutschland
Art des Produktes: Diverse Arten von Schulen, Spitalbauten, Kliniken, Notfallstationen, Bürobauten, Industriebauten, Turnhallen, Motelbau, Supermarkets, Tankstellen

Beschreibung des Produktes

Galvanisierte Stahlkonstruktionen, Aluminium-Curtain-Walls und Rasterwände, Metalldach- und Unterdeckenkonstruktion, bewegliche Trennwände, Sanitärpaneele

Produktion 1970

Montierte Gebäude pro Jahr: 100 000 m²
Verkaufspreis des fertigen Baus pro m²: 50 000–150 000 Lire
Gesamtumsatz pro Jahr: 10 000 000 000 Lire
Gesamte Unkosten für Arbeitskräfte pro Jahr (inklusive Beratungshonorare): 2 500 000 000 Lire
Gesamtwert des investierten Kapitals (Produktionsanlagen, Betriebsausrüstung, andere Anlagen): 5 000 000 000 Lire

Personal

Leitung: 10 Personen
Entwerfer und Techniker: 50 Personen
Technisches Personal für die Produktion: 25 Personen
Fabrikarbeiter: 350 Personen
Büroangestellte: 50 Personen
Personal für Transport und Versand: 5 Personen
Verkauf: 10 Personen
Montagearbeiter auf der Baustelle: 250 Personen
Andere: –
Gesamtbestand des Personals: 750 Personen

Arbeitsweise

Wer entwirft die Komponenten? IPI
Wer entwirft die Gebäude? IPI
Wer produziert die Komponenten? IPI
Wer finanziert die Gebäude? IPI
Wer montiert die Gebäude? IPI
Wer sucht die Kunden aus? IPI
Wer wählt die Komponentenproduzenten? –

Organization in operation since: 1942
Organization operating in: Italy, Greece, Germany
Line of product: All types of school buildings, hospitals, clinics, ambulatories, dispensaries, office-buildings, factories, gymnasiums, motels, supermarkets, service stations

Description of product

Galvanized steel constructions, aluminium curtain- and grid-walls, metal roofing and ceilings, movable partition walls, sanitary panels

Production 1970

Sq.m of assembled building per year: 100 000
Sales price of sq.m of finished building: 50 000–150 000 Liras
Total value of sales per year: 10 000 000 000 Liras
Total cost of manpower per year (inclusive of consultants fees): 2 500 000 000 Liras
Total value of capital investment (production facilities, production equipment, other facilities): 5 000 000 000 Liras

Staff

Management: 10 persons
Designers and technical staff: 50 persons
Production technical staff: 25 persons
Factory workers: 350 persons
Clerical staff: 50 persons
Transportation and/or shipment staff: 5 persons
Sales force: 10 persons
Site assembly crew: 250 persons
Others: –
Total of staff employed or dependent on the organization: 750 persons

Operational way

Who designs the components? IPI
Who designs the buildings? IPI
Who produces the components? IPI
Who finances the building? IPI
Who assembles the building? IPI
Who does the marketing? IPI
Who selects the components manufacturers? –

1, 2 Schule in Mailand.
3 Schule in Deutschland.
4 «Zappatoni»-Spital, Florenz.
5 Turnhalle im Bau, Legnano.
6–8 Schule in Monopoli.

1, 2 School in Milan.
3 School in Germany.
4 "Zappatoni" Hospital, Florence.
5 Gymnasium under construction, Legnano.
6–8 School in Monopoli.

Forschung Research	Entwicklung Development	Landerwerb und -erschließung Land acquisition and land improvement	Bauentwurf Building design	Promotion Promotion	Marktforschung Marketing	Behördliche Dienste Bureaucratic service	Finanzierung Financing	Produktionsplanung Production scheduling	Produktion Production	Montageplanung Assembly planning	Montage Assembly	Wartungsdienst After-sale service
○	●	○	●	○	●	○	◍	●	●	●	●	●

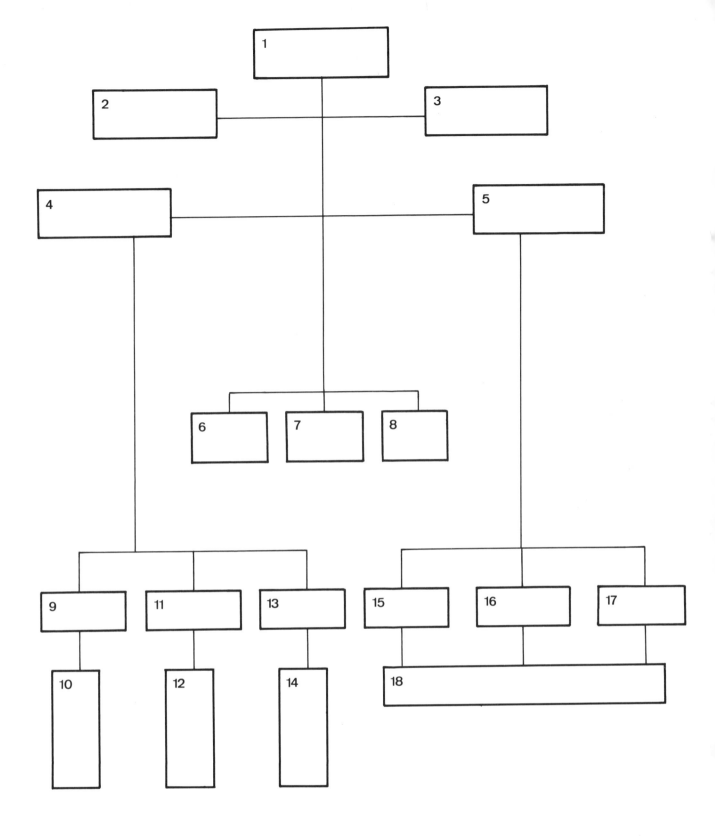

1 Präsident	1 President
2 Sekretariat	2 Secretary
3 Rechtsabteilung und Public Relations	3 Legal department and public relations
4 Technische Leitung	4 Technical management
5 Administrationsleitung	5 Administrative management
6 Verkauf Norditalien	6 Sales North Italy
7 Verkauf Süditalien	7 Sales South Italy
8 Verkauf Ausland	8 Foreign sales
9 Planung	9 Planning
10 Entwerfer und technisches Personal	10 Designers and technical staff
11 Produktion	11 Production
12 Produktion technisches Personal	12 Production technical staff
13 Bau	13 Building
14 Montagearbeiter auf Baustelle	14 Site assembly crew
15 Buchführung	15 Accounting
16 Einkauf	16 Buying
17 Versand	17 Shipping
18 Büropersonal	18 Clerical staff

Impresa Generale Costruzioni MBM Meregaglia S.p.A.

Via Rosselli, Trezzano sul Naviglio (Milan, Italy)

Organization in operation since: 1963
Organization operating in: Italy (Milan, Rome), U.S.A., France
Line of product: Any kind of building (residential, industrial, special), both conventionally and system-built using the Balency-MBM Industrialized Systems

Description of product

Balency-MBM Systems: load bearing concrete inside panels, load bearing concrete outside sandwich panels, cast in place (or precast), concrete slabs, all facilities incorporated in the panels and slabs

Production

Gross sq.m of assembled apartment buildings per year: 60 000–80 000
Sales price of gross sq.m of finished apartment building: 66 000 Liras
Total value sales per year: 6 000 000 000–7 000 000 000 Liras
Total cost of manpower per year (inclusive of consultants fees): 1 500 000 000 Liras
Total value of capital investment (production facilities, production equipment, other facilities): 2 000 000 000 Liras

Staff

Management: 3 persons
Designers and technical staff: 10 persons
Production technical staff: 7 persons
Factory workers: 110 persons
Clerical staff: 20 persons
Transportation and/or shipment staff: 8 persons
Sales force: 2 persons
Site assembly crew: 150 persons
Others: 90 persons
Total staff employed or dependent on the organization: 397 persons

Operational way

Who designs the components? Company's technical staff
Who designs the buildings? Company's technical staff with independent consultants architects
Who produces the components? Company's factory
Who finances the building? The company, local authorities, government, insurance companies, privates, others
Who assembles the buildings? Company's assembly crew
Who does the marketing? Company's staff
Who selects the components manufacturers? Company's staff

1, 2 Production line.
3 Bank building, Milan.
4 AGIP-Motel, San Donato, Milan.
5 Apartment houses, Trezzano sul Naviglio, Milan.
6 Apartment houses, quartiere Bovisasca, Milan.
7 Apartment houses, quartiere Gallaratese, Milan.
8 Assembly of panels.

89

91

8

Forschung Research	Entwicklung Development	Landerwerb und -erschließung Land acquisition and land improvement	Bauentwurf Building design	Promotion Promotion	Marktforschung Marketing	Behördliche Dienste Bureaucratic service	Finanzierung Financing	Produktionsplanung Production scheduling	Produktion Production	Montageplanung Assembly planning	Montage Assembly	Wartungsdienst After-sale service

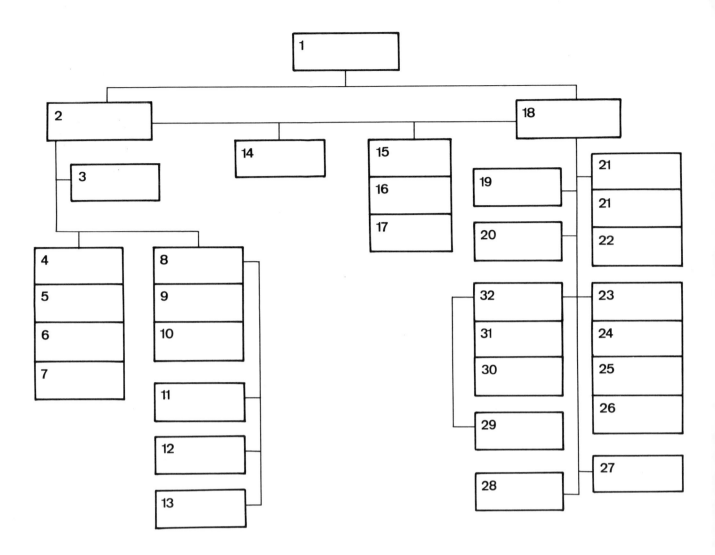

1 Generaldirektor	1 Managing director
2 Produktionsdirektor	2 Production director
3 Produktionsausrüstung	3 Production equipment
4 Fabrik	4 Factory
5 Paneelproduktion	5 Panels production
6 Werkstatt	6 Workshop
7 Allgemeine Dienste	7 General services
8 Organisation auf der Baustelle	8 Site organization
9 Unterhalt	9 Maintenance service
10 Elektrische Ausrüstung	10 Electrical equipment
11 Fertigungsdienst	11 Finishes assistance-service
12 Baustellen	12 Sites
13 Baustellen	13 Sites
14 Projektverwaltung	14 Projects administration
15 Kostenberechnung der Projekte	15 Projects accountancy
16 Nachkalkulation	16 After-calculation
17 Angebote	17 Offers
18 Zentralleitung	18 Central direction
19 Kaufmännisches Büro	19 Commercial office
20 Sekretariat	20 General secretary
21 Administration	21 Administration
22 Löhne	22 Salaries
23 Technisches Büro	23 Technical office
24 Gebäude	24 Buildings
25 Formen	25 Moulds
26 Ausrüstung	26 Equipment
27 Einkaufsbüro	27 Buying office
28 Zentrum für Informationsverarbeitung	28 Information elaboration centre
29 Arbeitsvorbereitung	29 Methods studies
30 Industrielle Kostenberechnung	30 Industrial accountancy
31 Terminplanung	31 Scheduling
32 Organisationsbüro	32 Organization office

Balency & Schuhl

278bis, avenue Napoléon-Bonaparte, 92 Rueil-Malmaison, France

Organization in operation since: 1909
Organization operating in: France, United Kingdom, Italy, Ireland, United States, etc.
Line of product: Office and apartment buildings, public buildings, private homes with prefab methods employing large reinforced concrete slabs; issuance of licences, engineering, construction of prefabrication works and machinery

Description of product

Manufactured in fixed or mobile plants, the prefab components are transported and then assembled on the building site. As much of the installations as possible (electricity, plumbing, etc.) is incorporated in the components in the factory, where also the finish of the external wall is generally done

Production 1970

Sq.m of assembled building per year: 230 000
Sales price of sq.m of finished building: Between 600 and 1 000 francs
Total value of sales per year: 180 000 000 francs (Balency and subsidiaries, without taxes)
Total cost of manpower per year: 60 000 000 francs (inclusive of consultants' fees)
Total value of capital investment (production facilities, production equipment, other facilities): 36 000 000 francs

Staff

Management: 20 persons
Designers and technical staff: 80 persons
Production technical staff: 25 persons
Factory workers: 430 persons
Clerical staff: 150 persons
Transportation and/or shipment staff: –
Sales force: 25 persons
Site assembly crew: 1470 persons
Others: –
Total of staff employed or dependent on the organization: 2200 persons

Operational form

Who designs the components? Balency & Schuhl Office of Technical Studies
Who designs the buildings? Architects or Balency & Schuhl Office of Technical Studies directly
Who produces the components? Balency & Schuhl
Who finances the building? The clients or bank credit
Who assembles the building? Balency & Schuhl
Who does the marketing? Balency & Schuhl
Who selects the components manufacturers? Balency & Schuhl

Remarks

The enterprise likewise grants licences covering processes and guarantees complete technical assistance to licensees.

1–3 Apartment houses, London-Thamesmead.
4 Apartment houses, Paris-Bobigny.
5 Students' dormitory, Châtenay-Malabry.
6 Standard one-family house.
7 Apartment houses, Vitry-sur-Seine.

3

4

5

6

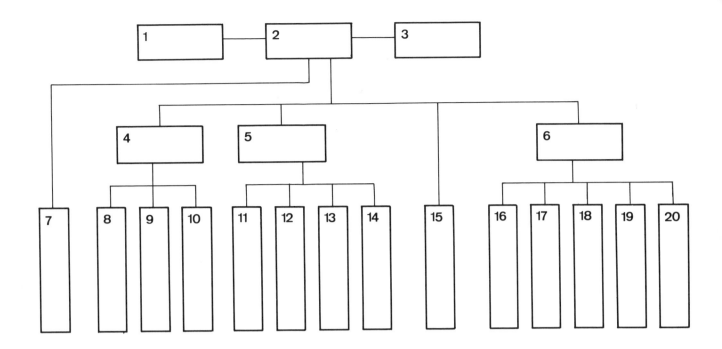

1 Präsident	1 President
2 Generaldirektor	2 General manager
3 Verwaltungsrat	3 Managing board
4 Generalsekretär	4 General secretary
5 Stellvertreter des Generaldirektors	5 Sub-manager
6 Stellvertreter des Generaldirektors	6 Sub-manager
7 Betriebskontrolle	7 Management supervision
8 Hauptbuchhaltung und analytische Buchhaltung	8 General and analytical accounting
9 Personalverwaltung	9 Personnel management
10 Material und Kontrolle der Anschaffungen	10 Materials and supplies
11 Agentur Perpignan	11 Perpignan Agency
12 Agentur Le Havre	12 Le Havre Agency
13 Autonome Abteilung für Frankreich	13 Independent works, France
14 Agentur Paris	14 Paris Agency
15 Technisches Büro Inland	15 National office of technical studies
16 Agentur Paris, Hochhäuser	16 Paris Agency, apartment houses
17 Bauprozesse und Lizenzen	17 Processes and licences
18 Autonome Abteilung Ausland	18 Independent works, abroad
19 Forschung und Entwicklung	19 Research and development
20 Filialen im Ausland	20 Subsidiaries abroad

Forschung / Research
Entwicklung / Development
Landerwerb und -erschließung / Land acquisition and land improvement
Bauentwurf / Building design
Promotion / Promotion
Marktforschung / Marketing
Behördliche Dienste / Bureaucratic service
Finanzierung / Financing
Produktionsplanung / Production scheduling
Produktion / Production
Montageplanung / Assembly planning
Montage / Assembly
Wartungsdienst / After-sale service

Constructions Modulaires S.A.

100, rue de Sèvres, 92 Boulogne-Billancourt, Frankreich

Tätig seit: 1966
Tätig in: Frankreich
Art des Produktes: Vorgefertigtes Bauen – System CLASP

Beschreibung des Produktes

Gelenkige Stahlstrukturen, Beton-Fassaden-Paneele, verschiedenartig vorgefertigte Paneele

Produktion 1970

Montierte Gebäude pro Jahr: 120 000 m^2
Durchschnittlicher Verkaufspreis pro m^2: Von 560 Frs. bis 650 Frs.
Gesamtumsatz pro Jahr: 73 328 000 Frs.
Gesamte Unkosten für Arbeitskräfte pro Jahr (inklusive Beratungshonorare): 1 563 921 Frs.
Gesamtwert des investierten Kapitals (Produktionsanlagen, Betriebsausrüstung, andere Anlagen): 1 369 565 Frs.

Personal

Leitung: 3 Personen
Entwerfer und Techniker: 12 Personen
Computerabteilung: 6 Personen
Büropersonal: 2 Personen
Verkauf: 12 Personen
Montagemannschaft auf Baustelle: 4 Personen
Andere: 27 Personen
Gesamtbestand des Personals, angestellt oder abhängig von der Firma: 66 Personen

Arbeitsweise

Wer entwirft die Komponenten? Constructions Modulaires
Wer entwirft die Gebäude? Freigewählte Architekten oder Architekturabteilung der C.M.
Wer produziert die Komponenten? Hersteller, die von C.M. ausgewählt werden
Wer finanziert die Gebäude? Die Auftraggeber
Wer montiert die Gebäude? Generalunternehmer, ausgewählt von C.M. oder vom Auftraggeber
Wer sucht die Kunden aus? Constructions Modulaires
Wer wählt die Komponentenproduzenten? Constructions Modulaires

Constructions Modulaires S.A.

100, rue de Sèvres, 92 Boulogne-Billancourt, France

Organization in operation since: 1966
Organization operating in: France
Line of product: Prefabricated building – System CLASP

Description of product

Pin-jointed steel structure, concrete cladding components and different prefabricated panels

Production 1970

Sq.m of assembled building per year: 120 000
Sales price of sq.m of finished building: From 560 Frs. to 650 Frs.
Total value of sales per year: 73 328 000 Frs.
Total cost of manpower per year (inclusive of consultants' fees): 1 563 921 Frs.
Total value of capital investment (production facilities, production equipment, other facilities): 1 369 565 Frs.

Staff

Management: 3 persons
Designers and technical staff: 12 persons
Computer department: 6 persons
Clerical staff: 2 persons
Sales force: 12 persons
Site assembly crew: 4 persons
Others: 27 persons
Total of staff employed or dependent on the organization: 66 persons

Operational way

Who designs the components? Constructions Modulaires
Who designs the buildings? External architects or C.M. Architecture Department
Who produces the components? Manufacturers selected by C.M.
Who finances the building? The clients
Who assembles the building? General contractors selected by C.M. or by the client
Who does the marketing? Constructions Modulaires
Who selects the components manufacturers? Constructions Modulaires

---

1, 2 Primarschule in Thiais. Architekt: M. Homberg.
3, 4 Sekundarschule in Moirans. Architekt: Gillet.
5, 6 Sekundarschule in Montendre. Architekt: Keyte.
   7 Primarschule in Le Pecq. Architekten: Hummel und Mathe.

1, 2 Primary school, Thiais. Architect: M. Homberg.
3, 4 Secondary school, Moirans. Architect: Gillet.
5, 6 Secondary school, Montendre. Architect: Keyte.
   7 Primary school, Le Pecq. Architects: Hummel and Mathe.

101

5

6

Forschung Research	Entwicklung Development	Landerwerb und -erschließung Land acquisition and land improvement	Bauentwurf Building design	Promotion Promotion	Marktforschung Marketing	Behördliche Dienste Bureaucratic service	Finanzierung Financing	Produktionsplanung Production scheduling	Produktion Production	Montageplanung Assembly planning	Montage Assembly	Wartungsdienst After-sale service
○	●	○	◍	○	●	○	○	◍	○	●	○	◍

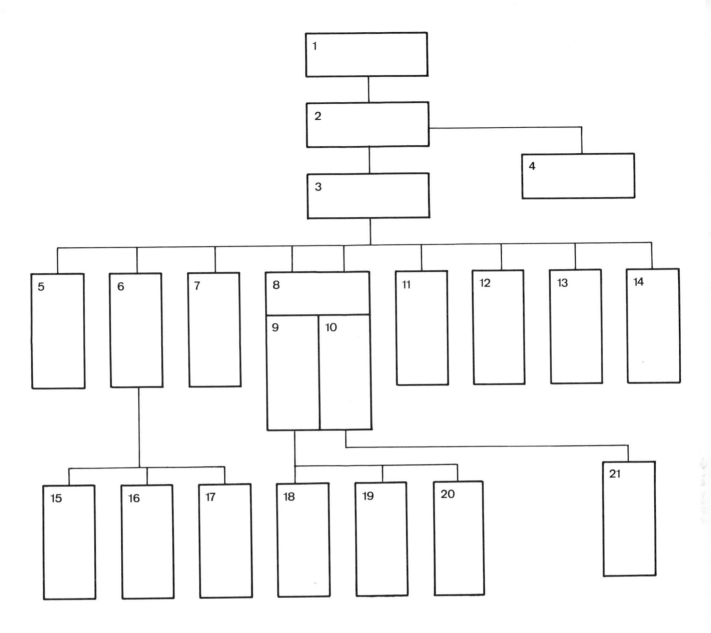

```
 1  Verwaltungsrat
 2  Generaldirektor – Präsident
 3  Generaldirektor
 4  Direktionssekretariat
 5  Verwaltung und Finanzen
 6  Projekte und Marktforschung
 7  Kaufmännische Dienste
 8  Bauleitung
 9  Technisches Büro
10  Verwaltung.
11  Public Relations
12  Computerabteilung
13  Forschungsabteilung
14  Abteilung Ausland
15  Architektur
16  Einschätzung – Verbesserung
17  Einschätzung – Beschreibung – Ausschreibung
18  Studienbüro
19  Projektleiter
20  Statische Abteilung
21  Einkauf
```

```
 1  Managing board
 2  President
 3  General manager
 4  Managing secretary
 5  Administration and finances
 6  Projects and marketing
 7  Commercial service
 8  Works department
 9  Technical division
10  Administrative
11  Public relations
12  Information
13  Research department
14  Foreign department
15  Architecture
16  Estimates – optimations
17  Estimates – descriptions – bidding
18  Study office
19  Project leaders
20  Structure studies
21  Purchasing service
```

Carl Möller

Osnabrück, German Federal Republic

Organization in operation since: 1865
Organization operating in: Northwest Germany
Line of Product: Building processes and components of all kinds for use in reinforced concrete constructions, bridge construction, industrial construction, prefab construction, finished construction for housing (multi-family houses, high risers), industrial and commercial buildings (factories, warehouses, sales premises, office buildings) and public buildings (communal buildings of all kinds, hospitals, schools, sports facilities, military barracks)

Description of product

No specialized programme, execution in accordance with individual requirements as well as plans of independent architects and engineers and plans of the firm's own construction division, elaboration of special proposals for rational programming and production following established types and systems

Production

Sq.m of assembled building per year: 30 000 sq.m
Average sales price per sq.m: DM 700.–
Total value of sales per year: 20 000 000 DM
Total labour costs per year: 6 500 000 DM
Total value of invested capital (offices, factories, equipment, etc.): 30 000 000 DM

Staff

Management: 4 persons
Technical staff: 41 persons
Production technical staff: 20 persons
Factory workers: 117 persons
Clerical staff: 28 persons
Transportation staff: 12 persons
Sales force: 2 persons
Site assembly crew: 197 persons
Others: 9 persons
Total employees: 430 persons

Operational form

Who designs the components? Firm's own construction office or architects and engineers as selected by client
Who designs the buildings? Firm's own design division or architects as selected by client
Who produces the components? Firm's own concrete component works or own contracting firm
Who finances the buildings? Banks or finance companies
Who assembles the buildings? Firm's own assembly division or sub-contractor
Who does the marketing? Firm's own canvassing division
Who selects the components manufacturers? Firm's own project office

1 Commercial building. Architect: Helmut Eggemann BDA.
2 High-risers, Belm-Powe. Architect: Interbau AG & Co. KG.
3 Sports arena. Architect: Grad. Eng. Werner Zobel.
4 Wholesale premises. Architect: Hans G. Garthaus.
5 High-rise office building. Architect: Prof. Fr. W. Kraemer.
6 Metal-working trade school, Osnabrück. Architects: Detschke, Däke, Simon.
7 Hospital, Harderberg. Architects: Wienker, Frense.

1

2

5

6

7

1	Akquisition
2	Geschäftsführung
3	Sekretariat
4	Kaufmännische Leitung
5	Technische Leitung: Innendienst
6	Technische Leitung: Außendienst
7	Mettinger Ziegelwerke
8	Finanzbuchhaltung/Gehaltsabrechnung
9	Lohnabrechnung
10	Einkauf
11	Betriebsbuchhaltung/Rechnungsprüfung
12	Betriebsabrechnung
13	Telefon/Empfang
14	Werkleitung
15	Oberbauleitung (Hochbau)
16	Technische Leitung (Tiefbau)
17	Schreibbüro
18	Produktion
19	Montage
20	Baustoffprüfstelle
21	Bauleitungen
22	Eisenbiegerei
23	Schreibbüro
24	Betriebsbüro
25	Bauleitung
26	Arbeitsbüro
27	Einsatzleitung
28	Werkleitung
29	Verkauf
30	Betriebsbüro
31	Einsatzleitung
32	Materialverwaltung/Magazin
33	Tischlerei/Zimmerei
34	Malerwerkstatt
35	Geräte und Fahrzeuginspektion
36	Maschinenwerkstatt/Schlosserei
37	Elektrowerkstatt
38	Schlüsselfertiges Bauen
39	Kalkulation
40	Planung und Terminierung
41	Projektbearbeitung
42	Kalkulation
43	Konstruktionsbüro
44	Planung und Terminierung

1	Acquisition
2	General management
3	Secretariat
4	Commercial management
5	Technical management: internal
6	Technical management: field service
7	Mettinger brick works
8	Financial accounting/salaries
9	Payrolls
10	Purchasing
11	Operations accounting/invoicing
12	General accountancy
13	Telephone/reception
14	Production management
15	Construction management (buildings, etc.)
16	Technical management (foundations, etc.)
17	Clerical office
18	Production
19	Assembly
20	Building-materials testing
21	Building-site management
22	Metal-working shop
23	Clerical office
24	Operations office
25	Building-site management
26	Project office
27	Coordination
28	Works management
29	Sales
30	Operations office
31	Coordination
32	Materials management/warehouse
33	Joinery/woodworking shop
34	Paint shop
35	Equipment and vehicle inspection
36	Machine shop/precision mechanics
37	Electrical shop
38	Finished building
39	Pricing
40	Planning and scheduling
41	Project readying
42	Pricing
43	Construction office
44	Planning and scheduling

Forschung / Research · Entwicklung / Development · Landerwerb und -erschließung / Land acquisition and land improvement · Bauentwurf / Building design · Promotion / Promotion · Marktforschung / Marketing · Behördliche Dienste / Bureaucratic service · Finanzierung / Financing · Produktionsplanung / Production scheduling · Produktion / Production · Montageplanung / Assembly planning · Montage / Assembly · Wartungsdienst / After-sale service

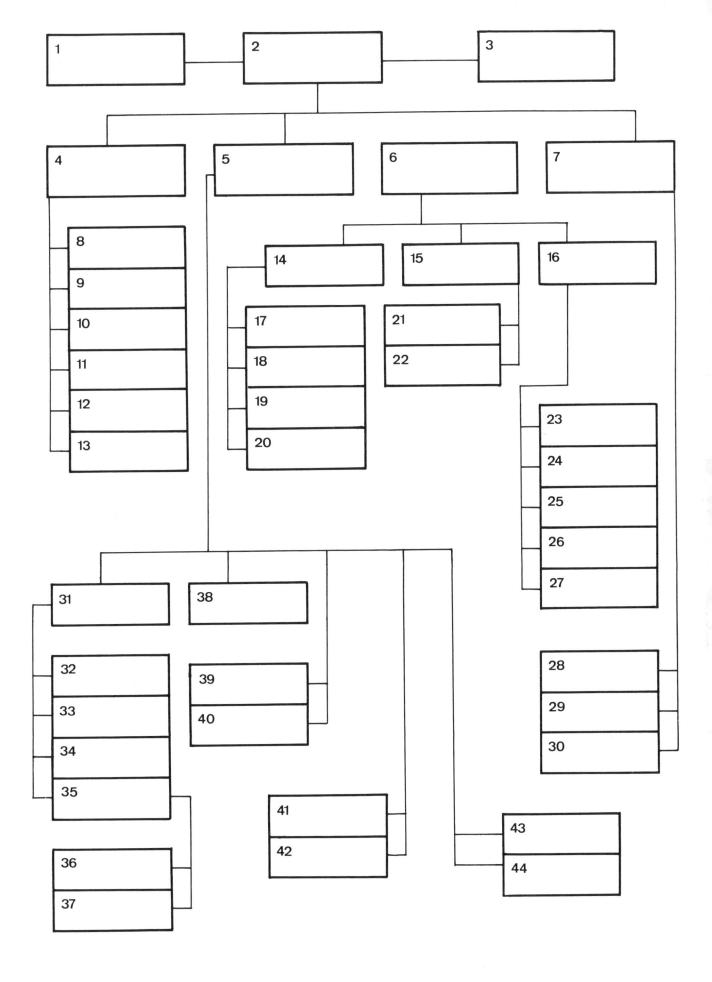

Nachbarschulte & Co. KG

4225 Gahlen, Rhld., Bundesrepublik Deutschland

Tätig seit: 1953
Tätig in: 200 km um den Hauptsitz, mit einigen Lizenznehmern in Nord- und Süddeutschland
Art des Produktes: Wohnhäuser, Wohnheime, Schulen, Kindergärten, Büros, Industriebauten, Sporthallen

Beschreibung des Produktes

Punktfundamente mit Stahlrahmen. Dach- und Deckenplatte als belüftetes Kaltdach in Leichtbauweise, freitragend. Die verschiedenen Tafeln und Platten werden untereinander mit Stahlteilen zug- und druckfest zu einem transportablen Raumelement verbunden

Produktion 1970

Gebaute Gebäude pro Jahr: 15 000 m²
Durchschnittlicher Verkaufspreis des fertigen Baus pro m²: DM 650.–
Gesamtumsatz pro Jahr: 10 000 000 DM
Gesamte Unkosten für Arbeitskräfte pro Jahr (inklusive Beratungshonorare): 1 300 000 DM
Gesamtwert des investierten Kapitals (Produktionsanlagen, Betriebsausrüstung, andere Anlagen): 1 800 000 DM

Personal

Leitung: 2 Personen
Entwerfer und technisches Personal: 5 Personen
Technisches Personal für Produktion: 10 Personen
Werkstattarbeiter: 65 Personen
Büropersonal: 11 Personen
Transport und/oder Versand: 4 Personen
Verkauf: 4 Personen
Montagemannschaft auf Baustelle: 22 Personen
Andere: 5 Personen
Gesamtbestand des Personals, angestellt oder abhängig von der Firma: 128 Personen

Arbeitsweise

Wer entwirft die Komponenten? Eigene Techniker
Wer entwirft die Gebäude? Teils eigene Architekten, teils fremde Architekten
Wer produziert die Komponenten? Überwiegend selbst
Wer finanziert die Gebäude? Diverse Kreditinstitute, Beratung durch Verkäufer
Wer montiert die Gebäude? Diverse Subunternehmer
Wer sucht die Kunden aus? Verkaufsabteilung
Wer wählt die Komponentenproduzenten? –

Nachbarschulte & Co. KG

4225 Gahlen, Rhld., German Federal Republic

Organization in operation since: 1953
Organization operating in: 200 km around the head office, with a few licences in northern and southern Germany
Line of product: Homes, residences, schools, kindergartens, office buildings, industrial buildings, sports facilities

Description of product

Point foundations and steel-frames. Roof and ceiling decks in light, cold-roof construction, freely supporting. The different panel and deck elements are tied in together by means of steel parts so to become undeformable and to constitute a transportable spatial element

Production 1970

Sq.m of assembled building per year: 15 000
Average sales price of finished construction per sq.m: DM 650.–
Total value of sales per year: 10 000 000 DM
Total labour costs per year (including consultants' fees): 1 300 000 DM
Total value of invested capital (offices, factories, equipment, etc.): 1 800 000 DM

Staff

Management: 2 persons
Design and technical staff: 5 persons
Production technical staff: 10 persons
Factory workers: 65 persons
Clerical staff: 11 persons
Transportation and/or shipping staff: 4 persons
Sales force: 4 persons
Site assembly crew: 22 persons
Others: 5 persons
Total staff, employed by or dependent on the firm: 128 persons

Operational form

Who designs the components? Firm's own technicians
Who designs the buildings? Partly firm's own architects, partly independent architects
Who produces the components? Mainly the firm
Who finances the buildings? Various banks, the sales representatives advise the potential buyer.
Who assembles the buildings? Various sub-contractors
Who does the marketing? Sales division
Who selects the components manufacturers? –

1–4 Schulbau aus standardisierten Einheiten.
  5 Bank.
6, 8 Einfamilienhaus. Architekt: Wolfgang Döring.
  7 Transportschema für die dreidimensionalen Einheiten.

1–4 School building of standardized units.
  5 Bank.
6, 8 One-family house. Architect: Wolfgang Döring.
  7 Transportation diagram for the three-dimensional units.

5

6

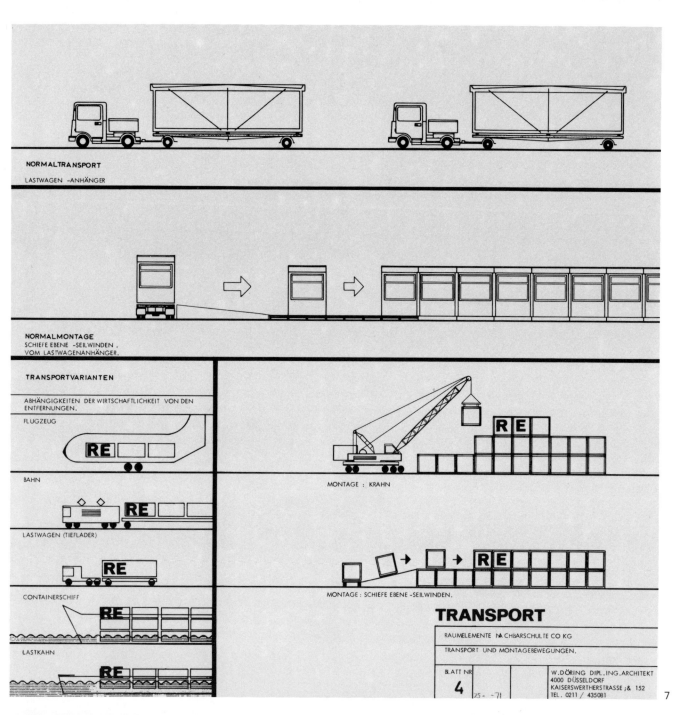

116

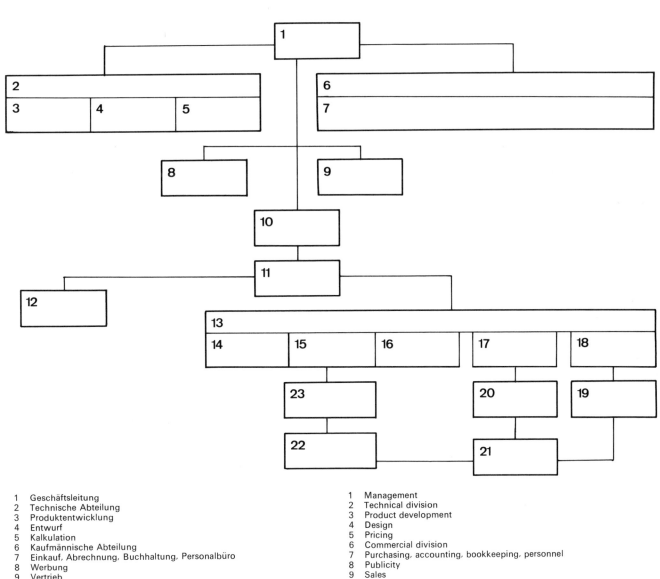

1 Geschäftsleitung	1 Management
2 Technische Abteilung	2 Technical division
3 Produktentwicklung	3 Product development
4 Entwurf	4 Design
5 Kalkulation	5 Pricing
6 Kaufmännische Abteilung	6 Commercial division
7 Einkauf, Abrechnung, Buchhaltung, Personalbüro	7 Purchasing, accounting, bookkeeping, personnel
8 Werbung	8 Publicity
9 Vertrieb	9 Sales
10 Betrieb	10 Operations
11 Auftragsannahme	11 Orders
12 Einholen der Baugenehmigung	12 Building permits
13 Arbeitsvorbereitung	13 Planning and scheduling
14 Schreinerei	14 Woodworking shop
15 Leimerei	15 Gluing shop
16 Schlosserei	16 Precision mechanics
17 Auswahl und Disposition/Subunternehmer für Ausbau	17 Selection and allocations/Sub-contractor for finishing
18 Bauleitung	18 Site supervisors
19 Fundamentierung	19 Foundation construction
20 Auswahl und Disposition/Subunternehmer für Fundamentierung	20 Selection and allocations/Sub-contractor for foundation construction
21 Baustellenmontage der Raumelemente	21 Building site assembly of spatial components
22 Transport der Raumelemente	22 Transportation of the spatial components
23 Tafelfertigung, Werkmontage der Raumelemente	23 Panel production, works assembly of the spatial components

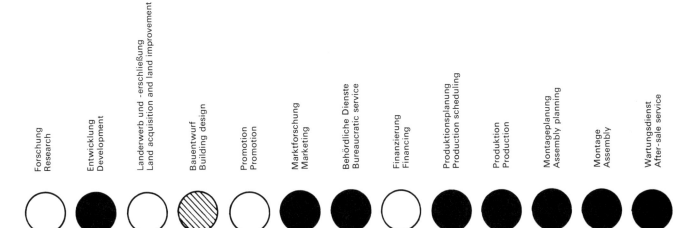

IGECO

3008 Bern, Könizstraße 74, Schweiz

Tätig seit: 1956
Tätig in: Schweiz; technische Beratung: Holland, Italien, Österreich
Art des Produktes: Entwürfe, Herstellung, Transport und Montage von vorgefertigten Elementen aus armiertem Beton und Spannbeton für den Wohnungsbau, Verwaltungsgebäude, Schulen, Industriebauten und Tiefbau, einschließlich Brücken

Beschreibung des Produktes

Gebäude: Die Vorfertigung IGECO besteht in der Werkausführung von Elementen (Tafeln) aus Stahlbeton (Wände, Fassaden, Deckenplatten usw.), deren Dimensionen einem Zimmer entsprechen. Auf den Baustellen charakterisiert sich das System IGECO durch die Trockenmontage der Bauelemente mit Schweißverbindungen (die dazu notwendigen Schweißplatten sind im Element eingegossen). Durch nachträgliches Schließen der Fugen mit Beton wird eine monolithische Struktur erzeugt. Industriebauten und Tiefbaukonstruktionen: Die IGECO-Fabriken verfügen über eine große Anzahl normierter Metallformen, über Dimensionierungsprogramme und Verbindungsdetails der Montage, die auf diese vorgefertigten Elemente abgestimmt sind

Produktion

Gesamtproduktion im Jahr: Vorgefertigte Betonelemente: 85 000 m^3 (212 500 t pro Jahr). Aufteilung: a) Wohnungsbau (ca. 75%): 65 000 m^3 (162 500 t pro Jahr), entspricht 1800 Wohnungen mit einer Baufläche von ca. 90 m^2. b) Industriebauten und Tiefbaukonstruktionen (ca. 25%): 20 000 m^3 (50 000 t pro Jahr)
Gesamtumsatz pro Jahr: 50 000 000 sFr.
Gesamte Lohn- und Gehaltskosten im Jahr: 11 500 000 sFr.
Gesamtwert des investierten Kapitals: 30 000 000 sFr. Wert der Investitionen für drei Werkanlagen (Bau, Ausrüstung für die Produktion und Montage, Werkzeuge usw., Boden nicht eingeschlossen)

Personal

Ingenieure, Techniker und Zeichner: 40 Personen
Büroangestellte und Verwaltungspersonal: 25 Personen
Technisches Personal für die Herstellung und Montage (Techniker, Werkführer, Vorarbeiter): 50 Personen
Qualifizierte und nicht qualifizierte Arbeitskräfte: 350 Personen (Herstellung), 130 Personen (Baustellenmontage)
Gesamtbestand des Personals: 600 Personen

Arbeitsweise

Wer entwirft die Komponenten? Technische Abteilung der IGECO oder Architekten und Ingenieure nach Wahl des Bauherrn
Wer entwirft die Gebäude? Außenstehende Architekten (IGECO stellt technische Unterlagen zur Verfügung)
Wer produziert die Komponenten? IGECO
Wer finanziert die Gebäude? Bauherren
Wer montiert die Gebäude? Montagemannschaft IGECO

1 Wohnhäuser, Adlikon-Regensdorf.
2 Brücke auf der Autobahn Lausanne–Genf.
3 Lagerplatz der Werkanlage Etoy.
4 Ein vorgefertigter standardisierter Balken.
5 Regalstapel-Lager
6 Wohnhäuser in Nyon.

IGECO

3008 Berne, Könizstrasse 74, Switzerland

Organization in operation since: 1956
Organization operating in: Switzerland. Technical assistance: Holland, Italy, Austria
Line of product: Studies, fabrication, transportation and assembly of prefab components of reinforced concrete and pre-stressed concrete intended for buildings, e.g., homes, villas, office buildings, schools, etc. and for industrial constructions and structural engineering projects, including bridges

Description of product

Buildings: The prefabrication processes of IGECO consist in fabricating at fixed works panels of reinforced concrete (walls, façades decks, etc.) whose dimensions correspond to a living unit (room). On the building sites, the IGECO system calls for dry assembly of the components, union by welding (metal elements incorporated) and the creation of a monolithic effect in the whole complex by concreting the joints. Industrial constructions and structural engineering projects: The IGECO factories possess a large number of standardized metal moulds, computerized dimensioning programmes and assembly node information relating to the prefab components corresponding to these moulds

Production

Total production per year: Prefab concrete components: 85 000 cu.m per year (212 500 tons per year). As follows: a) Buildings (approx. 75%): 65 000 cu.m per year (162 500 tons per year), equivalent to 1800 housing units with approx. 90 m^2 built-over surface. b) Industrial constructions and structural engineering projects (approx. 25%): 20 000 cu.m per year (50 000 tons per year)
Total value of sales per year: 50 000 000 SFr.
Total cost of manpower per year (social charges not included): 11 500 000 SFr.: amount of gross salaries and wages paid out to office staff and workers
Total value of capital investment: 30 000 000 SFr. Total value of investments for 3 factories (constructions, production and assembly equipment, apparatus, etc., grounds not included)

Staff

Engineers, technicians and designers: 40 persons
Clerical and administrative employees: 25 persons
Production technical staff and assembly crew (technicians, foremen, supervisors): 50 persons
Skilled and unskilled workers: 350 persons (production), 130 persons (site assembly)
Total of employed staff: 600 persons

Operational form

Who designs the components? Technical Division of IGECO or architects and engineers as selected by clients
Who designs the buildings? Outside architects (IGECO may provide engineering studies)
Who produces the components? IGECO
Who finances the buildings? The clients
Who assembles the buildings? IGECO assembly division

1 Apartment houses, Adlikon-Regensdorf.
2 Bridge on the Lausanne–Geneva motorway.
3 Storage area of the Etoy-production unit.
4 A prefabricated standardized beam.
5 Silo towers.
6 Apartment building at Nyon.

4

5

121

6

Forschung Research	Entwicklung Development	Landerwerb und -erschließung Land acquisition and land improvement	Bauentwurf Building design	Promotion Promotion	Marktforschung Marketing	Behördliche Dienste Bureaucratic service	Finanzierung Financing	Produktionsplanung Production scheduling	Produktion Production	Montageplanung Assembly planning	Montage Assembly	Wartungsdienst After-sale service

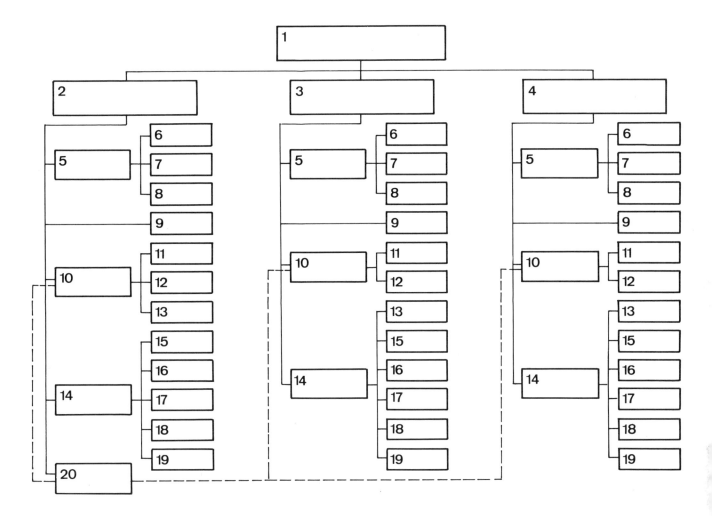

1	Generaldirektion Losinger	1 Losinger Company
2	IGECO SA, Etoy, Bevollmächtigter Verwalter	2 IGECO SA, Etoy, branch manager
3	IGECO AG, Lyßach, Bevollmächtigter Verwalter	3 IGECO AG, Lyssach, branch manager
4	IGECO AG, Volketswil, Bevollmächtigter Verwalter	4 IGECO AG, Volketswil, branch manager
5	Administration	5 Administration
6	Buchführung	6 Accounting
7	Sekretariat	7 Secretariat
8	Personal	8 Personnel
9	Verkauf	9 Sales
10	Technischer Dienst	10 Technical division
11	Geschäftsführung	11 Management
12	Technisches Büro	12 Technical office
13	Kontrolle	13 Inspection
14	Produktion	14 Production division
15	Planung	15 Planning
16	Arbeitsvorbereitung	16 Scheduling
17	Herstellung	17 Fabrication
18	Transport	18 Transportation
19	Montage	19 Assembly
20	Forschung und Auslandabteilung	20 Research and foreign division

Vic Hallam Limited

Langley Mill, Nottingham, England

Organization in operation since: 1921
Organization operating in: England and Holland
Line of product: System building for schools, offices, hospitals, housing and general amenity buildings

Description of product

Timber post, beam and panel systems for single- and two-storey work. Platform frame system (used mainly for housing) for three-storey work

Production 1970

Sq.m of assembled building per year: 325 000
Sales price of sq.m of finished building, superstructure: 20–25 £
Sales price of sq.m of finished building: 50–70 £
Total value of sales per year: £ 12 325 000
Total cost of manpower per year (inclusive of consultants' fees): £ 2 768 000
Total value of capital investment (production facilities, production equipment, other facilities): £ 2 803 471

Staff

Management: 170 persons
Designers and technical staff: 66 persons
Production technical staff: 107 persons
Factory workers: 1000 persons
Clerical staff: 260 persons
Transportation and/or shipment staff: 93 persons
Sales force: 22 persons
Site assembly crew: 307 persons
Others: 220 persons
Total of staff employed or dependent on the organization: 2245 persons

Operational way

Who designs the components? Vic Hallam design staff
Who designs the buildings? Mainly architects in private and public sector. Some buildings are designed by Vic Hallam staff
Who produces the components? Vic Hallam Ltd.
Who finances the building? Vic Hallam subsidiary company
Who does the marketing? Vic Hallam staff
Who selects the components manufacturers? Majority of components manufactured by Vic Hallam; other components selected by Vic Hallam or the client's architect

1 Secondary school at Cheddar.
2 Offices at Redditch.
3 Platform-frame housing at Marlpool, Derbyshire.
4 Llanwrst secondary school.
5 Computer centre for B.I.C.C.
6 Vic Hallam Ltd., factory, Langley Mill.
7 Motorway restaurant.

125

3

4

5

6

#	German	#	English
1	Präsident	1	Chairman
2	Direktor	2	Managing director
3	Stellvertretender Direktor	3	Deputy managing director
4	Verwaltungsrat	4	Board of directors
5	Tochtergesellschaft	5	Subsidiary company
6	Direktor für Baustellenmontage	6	Director site assembly
7	Inspektor des Fundamentes	7	Sub-structure inspector
8	Rohbaumontage	8	Superstructure erection
9	Dachbau	9	Roofing
10	Außenfertigung	10	External finishes
11	Glasarbeiten	11	Glazing
12	Innenfertigung	12	Internal finishes
13	Inspektion	13	Inspection
14	Unterhalt	14	Maintenance
15	Arbeitsplan	15	Work study
16	Kostenberechnung	16	Costing
17	Transport	17	Transport
18	Marktforschungsdirektor	18	Director of marketing
19	Marktforschung	19	Market research
20	Werbung	20	Publicity
21	Einschätzung	21	Estimating
22	Verkauf	22	Sales
23	Public Relations	23	Public relations
24	Kalkulation	24	Cost planning
25	Standards	25	Standards
26	Entwurf	26	Design
27	Entwicklung	27	Development
28	Forschung	28	Research
29	Produktionschef	29	Director of manufacture
30	Testen	30	Testing
31	Kostenberechnung	31	Costing
32	Arbeitsplan	32	Work study
33	Lager	33	Storage
34	Malerarbeiten	34	Painting
35	Fabrikanten	35	Fabrication
36	Formgebung	36	Machining
37	Einstellung der Einrichtungen	37	Setting out
38	Materialbearbeitung	38	Material handling
39	Trocknerei	39	Kilning
40	Einkauf	40	Purchasing

Forschung Research	Entwicklung Development	Landerwerb und -erschließung Land acquisition and land improvement	Bauentwurf Building design	Promotion Promotion	Marktforschung Marketing	Behördliche Dienste Bureaucratic service	Finanzierung Financing	Produktionsplanung Production scheduling	Produktion Production	Montageplanung Assembly planning	Montage Assembly	Wartungsdienst After-sale service
○	●	○	◐	○	●	○	○	●	●	●	◐	◐

7

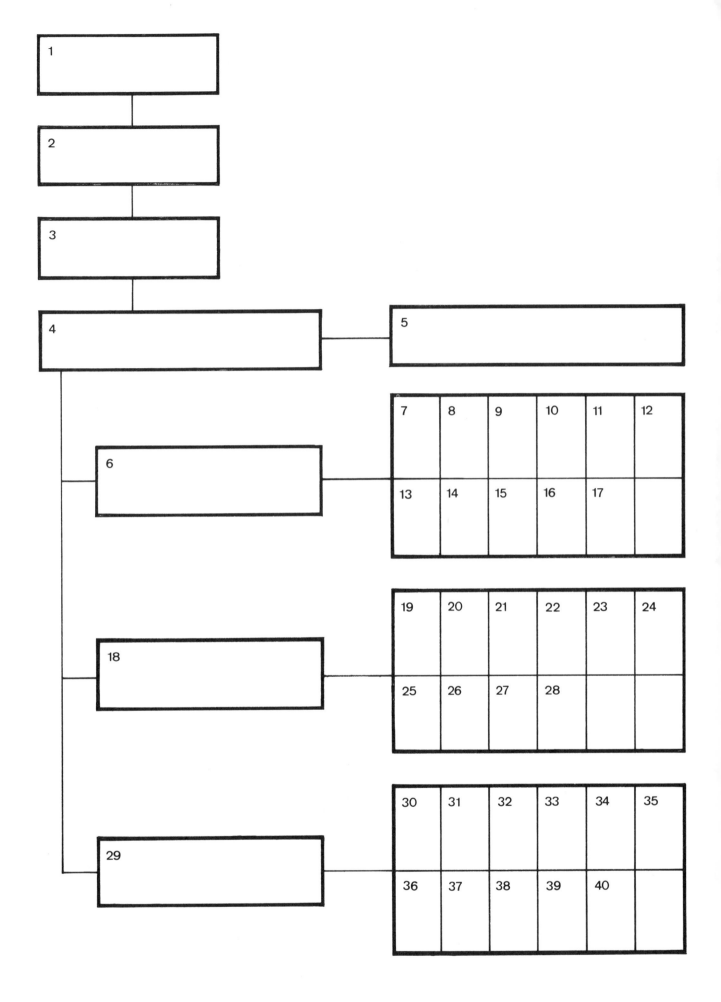

South Eastern Architect Collaboration (SEAC)

SEAC Central Development Group,
144A High Street, Epping, Essex, England

Organization in operation since: 1963
Organization operating in: South-East England, although it is possible to build anywhere in the British Isles
Line of product: Educational facilities and occasionally public facilities and office buildings

Description of product

Modular system based on a steel structure and different kinds of cladding

Production

Sq.m of assembled building per year: 250 000
Sales price of sq.m of finished building: £ 72.00
Total value of sales per year: £ 18 000 000
Total cost of manpower per year: Unknown but less than professional fees which would normally be paid
Total value of capital investment: Not known

Staff

Management: The chief architects of all members have overall management
Designers and technical staff: Employed by chief architects
Production technical staff: Employed by chief architects
Factory workers: Employed by sub-contractors
Clerical staff: Employed by chief architects
Transportation and/or shipment staff: Employed by sub-contractors (manufacturers)
Sales force: None
Site assembly crew: Employed by sub-contractors (builders)
Others: None
Total of staff employed or dependent on the organization: Unknown

Operational way

Who designs the components? Sub-contractors
Who designs the buildings? Project architect
Who produces the components? Sub-contractors
Who finances the buildings? Members of SEAC
Who assembles the buildings? A contractor appointed for each project
Who does the marketing? None
Who selects the components manufacturers? SEAC management

Remarks

SEAC is a consortia of central and local authorities building for themselves. However, a member of the consortia – C.E.D. Building Services (a branch of British Steel Corporation) – operates a "package deal" selling commercially. The overall design of the system is the responsibility of the consortia who selects manufacturers for the production of the components

1 Primary school at Chatham. Architect: E. T. Ashley Smith.
2 Sir Charles Lucas Comprehensive School, Colchester. Architect: Ralph Crowe.
3 East Hertford College of Further Education. Architect: G. C. Fardell.
4 Alderman Blaxill secondary school, Colchester. Architects: Greenwood and Abercrombie.
5 St. Albans School of Art. Architect: G. C. Fardell.
6 Holland-on-Sea primary school. Architect: Ralph Crowe.
7 Walter de Merton primary school. Architect: G. C. Fardell.
8 Primary school at East Malling. Architect: E. T. Ashley Smith.

5

6

Forschung Research	Entwicklung Development	Landerwerb und -erschließung Land acquisition and land improvement	Bauentwurf Building design	Promotion Promotion	Marktforschung Marketing	Behördliche Dienste Bureaucratic service	Finanzierung Financing	Produktionsplanung Production scheduling	Produktion Production	Montageplanung Assembly planning	Montage Assembly	Wartungsdienst After-sale service

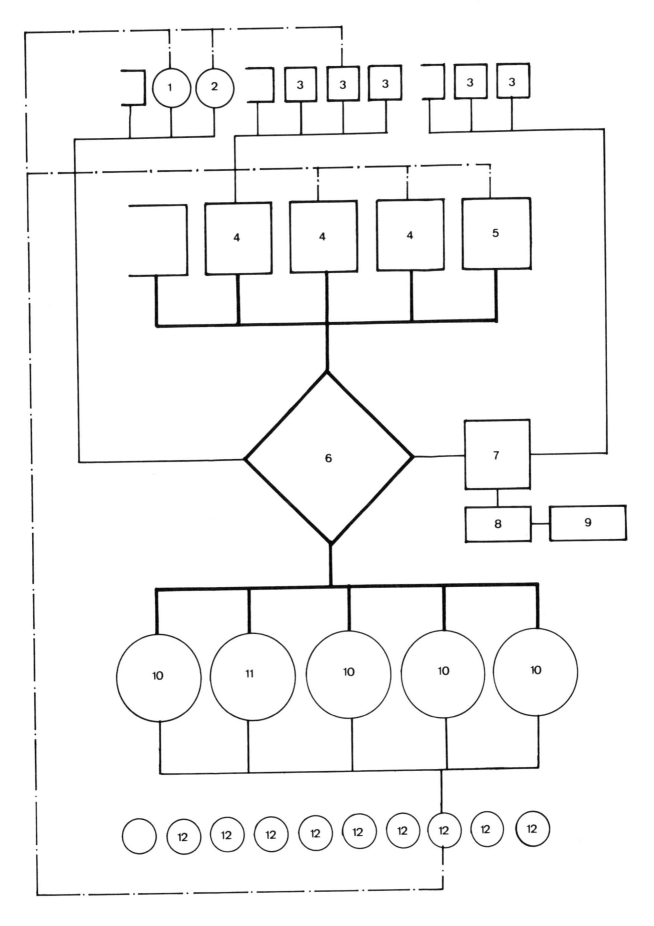

1 Kommerzielles Mitglied	1 Commercial member
2 Universität	2 University
3 Assoziiertes Mitglied	3 Associate member
4 Vollberechtigtes Mitglied	4 Full member
5 Umwelt-Ministerium	5 Department of environment
6 Zentrale Entwicklungsgruppe	6 Central development group
7 Ministerium für Erziehung und Wissenschaft	7 Department of education and science
8 Arbeitsgruppe für die technische Koordination	8 Technical coordinating working party
9 Andere Schulkonsortien	9 Other school consortia
10 Produzenten	10 Supplier
11 Produktion und Montage	11 Supply and fix
12 Generalunternehmer	12 General contractor

Kombinat für Wohnungsbau Nr. 3

Moskau, Domostroitelnaia 2, UdSSR

Tätig seit: 1965
Tätig in: Stadtgebiet von Moskau, manchmal auch in anderen Gebieten
Art des Produktes: Vorgefertigte Wohnungen aus Stahlbeton

Beschreibung des Produktes

Lasttragende Stahlbetonpaneele, produziert in einem kontinuierlichen Prozeß oder durch Standardformen. Bis 16geschossige Häuserblöcke. Betonböden und Dachplatten. Treppenhäuser aus Beton

Produktion 1971

Gebaute Gebäude pro Jahr: 500 000 m²
Durchschnittlicher Verkaufspreis pro m² eines fertigen Baus: 140 Rubel
Gesamtumsatz pro Jahr: 70 000 000 Rubel
Gesamte Unkosten für Arbeitskräfte pro Jahr (inklusive Beratungshonorare): 15 000 000; 8 400 000 für Fabrikarbeiter; 4 900 000 für Arbeiter auf der Baustelle
Gesamtwert des investierten Kapitals (Produktionsanlagen, Betriebsausrüstung, andere Anlagen): 15 000 000 Rubel

Personal

Leitung: 6 Personen
Entwerfer und technisches Personal: 70 Personen
Technisches Personal für Produktion: 60 Personen
Werkstattarbeiter: 1800 Personen
Büropersonal: 64 Personen
Transport und/oder Versand: Keine, auswärtige Arbeitskräfte
Verkäufer: –
Baustellenarbeiter: 2000 Personen
Andere: –
Gesamtbestand des Personals: 4000 Personen

Arbeitsweise

Wer entwirft die Komponenten? Die Komponenten werden vom technischen Personal des Kombinates nach allgemeinen sowjetischen Normen entworfen. Manchmal werden auch spezielle Komponenten für nichtnormierte Arbeiten angefertigt
Wer entwirft die Gebäude? Die Bauwerke werden vom lokal verantwortlichen Architekturinstitut nach den sowjetischen Normen entworfen
Wer produziert die Komponenten? Alle Betonkomponenten werden vom Kombinat produziert. Andere Komponenten werden auswärts gekauft
Wer finanziert die Gebäude? Der Staat, manchmal auch Staatsbanken für kooperative Bauten
Wer sucht die Kunden aus? Es gibt kein eigentliches Marketing, da die Produktion einen integrierten Bestandteil der gesamten Wirtschaftsplanung darstellt
Wer wählt die Komponentenproduzenten? Die Verantwortung für die Wahl liegt beim Kombinat
Wer montiert die Gebäude? Das Kombinat

1 Zwölfgeschossiger Prototyp, Moskau.
2 Montage eines Treppenelementes.
3–6 Container mit diversen Elementen.
7 Neungeschossiger Standardbau mit Balkonen.
8 Montage eines Bauwerkes.
9 Vorgefertigte Balkone.
10 21geschossiger Prototyp.
11 Ein Container wird beladen.

House-Building Combine No. 3

Domostroitelnaia 2, Moscow, U.S.S.R.

Organization in operation since: 1965
Organization operating in: Moscow urban area, occasionally outside
Line of product: Reinforced concrete prefabricated dwellings

Description of product

Load bearing reinforced concrete panels, produced by continuoud process or standard moulds. Housing blocks up to 16 floors. Concrete floor- and roof slabs. Concrete staircases

Production 1971

Gross sq.m of assembled apartment building per year: 500 000
Sales price of gross sq.m of finished building: 140 rubles
Total value of sales per year: 70 000 000 rubles
Total cost of manpower per year (inclusive of consultants' fees): 15 000 000 of rubles of which: 8 400 000 factory labour; 4 900 000 site labour
Total value of capital investment (production facilities, production equipment, other facilities) up to date: 15 000 000 rubles

Staff

Management: 6 persons
Designers and technical staff: 70 persons
Production technical staff: 60 persons
Factory workers: 1800 persons
Clerical staff: 64 persons
Transportation and/or shipment staff: None, outside labour
Sales force: None
Site assembly crew: 2000 persons
Others: –
Total of staff employed or dependent on the organization: 4000 persons

Operational way

Who designs the components? The components are designed by the technical staff of the Combine, according to the general Soviet Union standards. Special components for non-standard jobs do also occur
Who designs the buildings? The buildings are designed by the local responsible architectural institute, according to the general Soviet Union standards
Who produces the components? All concrete components are produced by the Combine itself. Other components are bought outside
Who finances the building? The State, occasionally state banks for cooperative buildings
Who does the marketing? No marketing activity as such exists, the production being an integrated part of the general development plan
Who selects the components manufacturers? The Combine itself is responsible for the selection
Who assembles the buildings? The Combine

1 12-stories prototype building, Moscow.
2 Installation of a staircase element.
3–6 Containers with various kind of elements.
7 9-stories standard building with balconies.
8 A building during erection.
9 Prefabricated balconies.
10 21-stories prototype building.
11 Loading of a container.

4

5

6

7

8

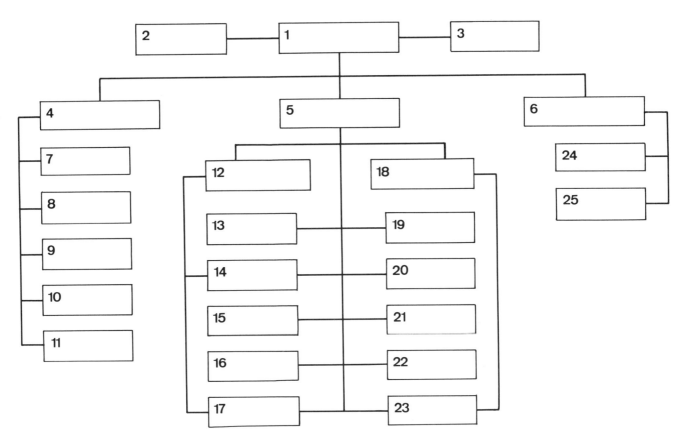

1 Direktor	1 Chief of the combine
2 Buchführung	2 Accountancy
3 Management	3 Management section
4 Vizedirektor für wirtschaftliche Fragen	4 Adjoint chief for economic questions
5 Chefingenieur	5 Chief engineer
6 Vizedirektor für Normierung und standardisierte Produktion	6 Adjoint chief for normalization and standard production
7 Planung	7 Planning
8 Lohn und Gehalt	8 Salary and wages
9 Forschung	9 Scientific research
10 Kostenberechnung und Verträge	10 Estimates and contracts
11 Wirtschaftliche Untersuchungen	11 Economical studies
12 Bauleiter	12 Chief of construction
13 Produktion	13 Production service
14 Versand	14 Dispatching
15 Technischer Dienst	15 Technical service
16 Leiter der Sicherheitskontrolle	16 Chief of safety control
17 Abteilung für Montage und Installation	17 Assembly and installation section
18 Produktionsleiter	18 Chief of production
19 Technologischer Dienst	19 Technological service
20 Mechanischer und elektrischer Dienst	20 Mechanic and power service
21 Produktionsabteilung	21 Production department
22 Versuchslaboratorium	22 Testing laboratory
23 Werkhalle Wostriakowski	23 Factory of Vostriakovsky
24 Transport	24 Transport service
25 Leitung der Normierung und Spezialherstellung	25 Direction of normalization and special production

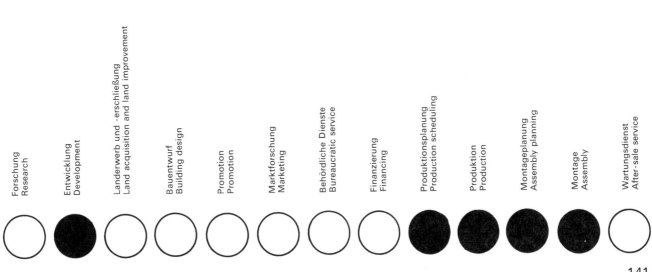

141

Kombinat für Wohnungsbau Nr. 1

Moskau, 2° Horoschewskij proeza 3, UdSSR

Integrated Home-Building Factory No. 1

2° Horoscevskij proeza, 3, Moscow, U.S.S.R.

Tätig seit: 1962 (in der heutigen Form)
Tätig in: Moskau, für einige Komponenten auch anderswo
Art des Produktes: Vollständige Wohnungen und einzelne Komponenten (sanitäre Zellen)

Beschreibung des Produktes

Stahlbetonbauten nach allgemeinen sowjetischen Normen (siehe Abbildungen von Kombinat Nr. 3). Ein spezielles Produkt des Werkes Nr. 1 ist eine vollständig eingerichtete sanitäre Zelle. Sie enthält getrenntes Badezimmer und Toilette und besteht aus vier Gips-Beton-Wänden, einer Trennwand und einem Stahlbetonboden mit Bodenbelag
Allgemeine Bemerkungen: Das Produkt gleicht dem von Kombinat Nr. 3, deshalb werden nur die wichtigsten Merkmale notiert

Produktion 1971

Montierte Bauten pro Jahr: 760 000 m²

Personal

Gesamtbestand des Personals: 8000 Personen

Arbeitsweise

Vier spezialisierte Produktionseinheiten und diverse Montagegruppen stehen unter einer Leitung. Diese Organisation ist momentan in der Sowjetunion die einzige ihrer Art und bietet wohl erhebliche Vorteile für die Montage auf der Baustelle (garantierter Versand und Terminkontrolle). Jedes Werk produziert jedoch auch Komponenten für Drittparteien und trägt die finanzielle Verantwortung

Organization in operation since: 1962 (under present form)
Organization operating in: Moscow and for some components outside
Line of product: Complete dwellings and individual components (sanitary boxes)

Description of product

Reinforced concrete dwellings according to the general Soviet Union standards (see pictures of Combine No. 3). A special product of the Factory No. 1 is the completely equipped sanitary box unit. The box unit incorporates isolated bathroom and water closet facilities and presents a thin walled spatial structure formed by four gypsum-concrete walls, a partition and a reinforced concrete bottom with flooring
General remarks: Due to the similarity of the product to the preceeding one (Combine No. 3) only some essential features are indicated in this case

Production 1971

Gross sq.m of assembled apartment building: 760 000

Staff

Total of staff employed: 8000 persons

Operational way

Relevant in this case is the integration of four specialized manufacturing units and various erection groups under one organizational roof. This approach is unique for the moment in the Soviet Union and it seems to offer substantial advantages for site construction (insured delivery and time control). Each factory, however, can and does produce components for third parties, and is financially responsible

1 Modell des Prototyps für die fertige sanitäre Zelle.
2 Werkhalle für die Herstellung der sanitären Zellen in Choroschewo. Der Armierungskäfig wird in die Form gesenkt (rechts), Ausrüstung der Zellen (links), ein Modell der fertigen Zelle (ganz rechts).

1 Prototype model of the finished sanitary box.
2 Factory hall for the production of sanitary cells at Khoroshevo. On the right, the reinforcement cage is lowered on the mould, on the left, ready boxes in the process of being equipped, on the extreme right, a model of the finished box.

Forschung Research	Entwicklung Development	Landerwerb und -erschließung Land acquisition and land improvement	Bauentwurf Building design	Promotion Promotion	Marktforschung Marketing	Behördliche Dienste Bureaucratic service	Finanzierung Financing	Produktionsplanung Production scheduling	Produktion Production	Montageplanung Assembly planning	Montage Assembly	Wartungsdienst After-sale service
○	●	○	○	○	○	○	○	●	●	●	●	○

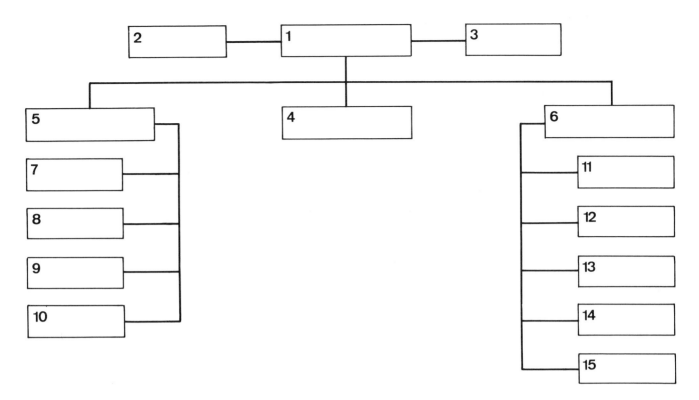

```
1       Leitung
2       Normierung und Forschung
3       Konstruktionslaboratorium
4       Leitung der Normierung
5       Produktion
6       Bauabteilung
7       Werkhalle Choroschewo (sanitäre Zellen)
8       Werkhalle Tuschino (Bodenpaneele und Wände im Erdgeschoß)
9       Werkhalle Krasnaja-Presnija (Außenwände und Dachpaneele)
10      Werkhalle Rostokino (lasttragende Paneele, Treppenabsätze und
        Treppenläufe, usw.)
11—15   Montageleitung
```

```
1       Direction
2       Normalization and research station
3       Construction laboratory
4       Direction of normalization
5       Production department
6       Construction department
7       Khoroshevo factory (sanitary boxes)
8       Tushino factory (floor panels and ground-floor walls)
9       Krasnaya-Presnya factory (external walls and roof panels)
10      Rostokino factory (load bearing panels, stair landings and flights, etc.)
11—15   Erection management
```

Servicio Técnico de Construcciones Modulares S.A.

Calle Angeles 3, Barcelona 1, Spanien

Servicio Técnico de Construcciones Modulares S.A.

Calle Angeles 3, Barcelona 1, Spain

Tätig seit: 1969
Tätig in: Spanien, Algerien
Art des Produktes: Ein modulares, industrialisiertes System für Schul- und Bürobauten

Beschreibung des Produktes

Leichte modulare Stahlstruktur mit Leichtbeton-Deckenplatten und Sandwich-Schwerbeton-Verkleidungspaneelen

Produktion 1971

Gebaute Gebäude pro Jahr: 20 000 m²
Durchschnittlicher Verkaufspreis eines fertigen Baus pro m²: 5000 Pesetas
Gesamtumsatz pro Jahr: 100 000 000 Pesetas
Gesamte Unkosten für Arbeitskräfte pro Jahr (inklusive Beratungshonorare): 7 000 000 Pesetas
Gesamtwert des investierten Kapitals (Büroeinrichtungen und Entwicklungskosten) bis jetzt: 4 500 000 Pesetas

Personal

Leitung: 3 Personen
Entwerfer und technisches Personal: 7 Personen
Technisches Personal für die Produktion: –
Werkstattarbeiter: –
Büropersonal: 7 Personen
Transport und/oder Versand: –
Verkauf: 2 Personen
Baustellenarbeiter: –
Andere: –
Gesamtbestand des Personals: 19 Personen

Arbeitsweise

Allgemein: Die Firma hat zuerst ein Bausystem entwickelt, das von der Behörde verwendet wird. Sie stellt den organisatorischen Rahmen in Zusammenarbeit mit unabhängigen Architekten und Bauunternehmern zur Verfügung
Marketing: Durch die Firma
Entwurf der Systeme und Komponenten: Durch die Firma
Bauentwurf: Firmeneigene Architekten und unabhängige Architekten
Werbung: Durch die Firma
Produktion der Komponenten: Eine ausgewählte Gruppe von spanischen Herstellern unter Jahresvertrag
Baustellenmontage: Die industrialisierten Elemente werden durch ihre Hersteller montiert, der übrige Teil des Baus durch lokale Bauunternehmer, die ad hoc ausgewählt werden
Vertragsabschlüsse: Durch die Firma selber oder einen ad hoc assoziierten Bauunternehmer

Organization in operation since: 1969
Organization operating in: Spain, Algeria
Line of product: A modular industrialized system for educational and office facilities

Description of product

A light modular steel structure with light concrete floor slabs and heavy concrete outside wall sandwich panels

Production 1971

Sq.m of assembled building per year: 20 000
Sales price of sq.m finished building: 5000 Pts./metre
Total value of sales per year: Pesetas 100 000 000
Total cost of manpower per year (inclusive of consultants' fees): 7 000 000 Pesetas
Total value of capital investment (office facilities and development cost) up to date: 4 500 000 Pesetas

Staff

Management: 3 persons
Designers and technical staff: 7 persons
Production technical staff: None
Factory workers: None
Clerical staff: 7 persons
Transportation and/or shipment staff: None
Sales force: 2 persons
Site assembly crew: None
Others: –
Total of staff employed or dependent on the organization: 19 persons

Way of operation

General: The company, having developed the technical system, promotes its use by governmental authorities or provides the organizational framework with the collaboration of independent architects and building contractors
Marketing: By the company
System and components design: By the company
Building design: Company architect or independent architect
Advertising: By the company
Components production: A selected ring of Spanish manufacturers under yearly contract
Site assembly: The industrialized components are assembled by the components manufacturers, the rest of the building by local contractors selected "ad hoc"
Contracting: By the company itself or by an "ad hoc" associated contractor

1 Schulhaus Can Clos in Barcelona.
2 Schulhaus Turo de la Peira in Barcelona.
3, 4 Hotel Ziri in Algier.
5 Treppenhaus im Bürogebäude Modulteu.
6 Bankgebäude in Barcelona.
7 Eingang zum Bürohaus Modulteu, Barcelona.

1 Can Clos school in Barcelona.
2 Turo de la Peira school in Barcelona.
3, 4 Hotel Ziri in Algiers.
5 Stairwell in the Modulteu Building.
6 Bank in Barcelona.
7 Entrance to Modulteu Building, Barcelona.

3

4

5

6

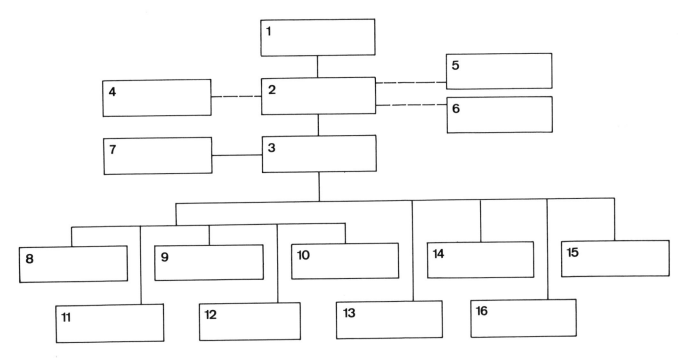

1 Direktionsvorstand
2 Generaldirektor
3 Geschäftsleiter
4 Technischer Berater
5 Buchhaltungsberater
6 Juristischer Berater
7 Sekretariat
8 Architekturabteilung
9 Statikabteilung
10 Entwicklung
11 Bauleitung und Arbeitsplanung
12 Arbeitspläne
13 Buchhaltung
14 Verkauf
15 Kalkulation
16 Akquisition

1 Board of directors
2 Managing director
3 General manager
4 Technical consultant
5 Accountancy consultant
6 Legal consultant
7 Secretary
8 Architecture department
9 Structure department
10 Development
11 Site supervision and planning
12 Working drawings
13 Accountancy
14 Sales
15 Costs calculations
16 Acquisition

Forschung / Research
Entwicklung / Development
Landerwerb und -erschließung / Land acquisition and land improvement
Bauentwurf / Building design
Promotion / Promotion
Marktforschung / Marketing
Behördliche Dienste / Bureaucratic service
Finanzierung / Financing
Produktionsplanung / Production scheduling
Produktion / Production
Montageplanung / Assembly planning
Montage / Assembly
Wartungsdienst / After-sale service

151

Staatliche Baugesellschaft Nr. 43

Budapest XI, Dombovári U. 17–19, Ungarn

State Building Organization No. 43

Budapest XI, Dombovári U. 17–19, Hungary

Tätig seit: Frühling 1948
Tätig in: Ungarn (hauptsächlich im Umkreis von Budapest)
Art des Produktes: Wohn- und Schulbauten

Beschreibung des Produktes

Betonsystem. Für den Wohnungsbau werden russische und dänische Lizenzen verwendet. Für den Schulhausbau bestehen eigene ungarische Entwicklungen. Die Gesellschaft produziert auch andere Gebäude wie Hotels, Spitäler, Lagerhäuser

Produktion 1971

Montierte Gebäude pro Jahr: 350 000 m² (nur Wohnbau)
Verkaufspreis des fertigen Hauses: 5431 Forint/m²
Gesamtwert der Verkäufe pro Jahr: 1 992 200 000 Forint
Gesamtkosten für Arbeitskräfte pro Jahr: 222 700 000 Forint (einschließlich Beratungshonorare)
Gesamtwert des investierten Kapitals: 924 800 000 Forint

Personal

Leitung und Büropersonal: 621 Personen
Entwerfer und Techniker: 820 Personen
Technisches Personal für die Produktion: 564 Personen (in den oben genannten 820 Personen inbegriffen)
Fabrikarbeiter: 1521 Personen
Transport und Versand: 359 Personen
Verkauf: Keine
Montagearbeiter auf der Baustelle: 155 Personen
Andere: 300 Personen
Gesamtbestand des direkt angestellten Personals: 1741 Personen (Leitung, Entwerfer, Techniker usw.)

Arbeitsweise

Wer entwirft die Komponenten? Eine staatseigene Firma
Wer entwirft die Gebäude? Eine staatseigene Firma
Wer produziert die Komponenten? Die drei Fabriken der staatlichen Baugesellschaft Nr. 43
Wer finanziert die Gebäude? Zu 90% der Stadtrat von Budapest
Wer sucht die Kunden aus? Keine eigentliche Marketing-Tätigkeit. Die Produktion wird von staatlichen Firmen besorgt
Wer wählt die Komponentenproduzenten aus? Die meisten Komponenten werden von der Gesellschaft selbst hergestellt
Wer montiert die Gebäude? Die Gesellschaft selbst

Organization in operation since: Spring 1948
Organization operating in: Hungary (mainly the Budapest area)
Line of product: Housing and educational buildings

Description of product

Concrete system. Russian and Danish licenses are utilized for the housing system. Original Hungarian development of the school building system. The organization also produces other kind of buildings like hotels, hospitals, warehouses

Production 1971

Gross sq.m of assembled building per year: 350 000 sq.m (housing only)
Sales price of gross sq.m of finished buildings: 5431 Florins/sq.m (exclusive of external connections to light, water, etc.)
Total value of sales per year: 1 992 200 000 of Florins
Total cost of manpower per year: 222 700 000 of Florins (inclusive of consultants' fees)
Total value of capital investment: 924 800 000 of Florins

Staff

Management and clerical staff: 621 persons
Designers and technical staff: 820 persons
Production technical staff: 564 persons (already included in the above 820)
Factory workers: 1521 persons
Transportation and shipment staff: 359 persons
Sales force: None
Site assembly crews: 155 persons
Others: 300 persons
Total of staff employed directly: 1741 persons (management, designers, technical staff and others)

Operational way

Who designs the components? State own design organization
Who designs the building? State own design organization
Who produces the components? The three factories of the State Building Organization No. 43
Who finances the buildings? 90% the Budapest Town Council
Who does the marketing? No marketing activity as such. The production is absorbed by the State organizations
Who selects the components manufacturers? Most components are produced by the Organization itself
Who assembles the buildings? The Organization itself

1 Wohnsiedlung «Óbuda».
2 Die Komponenten werden in Fabrik Nr. 1 verladen.
3 Fabrik Nr. 2.
4 Abtransport der Komponenten ab Fabrik Nr. 2.
5 Wohnsiedlung «Zugló».
6 Kindergarten in der Wohnsiedlung «Zugló».
7 Herstellung von Armierungsgittern in Fabrik Nr. 3.
8 Wohnsiedlung «Rákospalota».

1 Housing development "Óbuda".
2 Loading of components in Factory No. 1.
3 The Factory No. 2.
4 Shipment from Factory No. 2.
5 Housing development "Zugló".
6 Kindergarten in the "Zugló" housing development.
7 Preparation of steel-reinforcement in the Factory No. 3.
8 Housing development "Rákospalota".

153

3

4

5

6

7

8

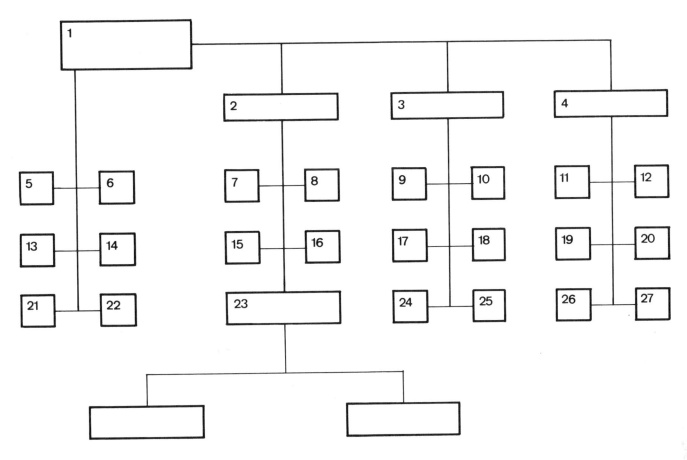

1	Generaldirektor	1 General manager
2	Produktions-Stellvertreter des Generaldirektors	2 Production sub-manager
3	Technischer Stellvertreter des Generaldirektors	3 Technical sub-manager
4	Wirtschaftlicher Stellvertreter des Generaldirektors	4 Financial sub-manager
5	Sekretariat	5 Secretariat
6	Technischer Berater	6 Technical consultant
7	Abteilung für Maschinenwesen	7 Mechanical department
8	Abteilung für Materialversorgung	8 Supply department
9	Bautenplanungsbüro	9 Construction planning office
10	Entwicklungen und Produktkonstruktionsbüro	10 Development and product construction office
11	Finanzabteilung	11 Finance department
12	Rechnungsabteilung	12 Accounting department
13	Rechtsabteilung	13 Legal department
14	Personalabteilung	14 Personnel department
15	Hauptenergetiker	15 Chief engineer
16	Haupttechnologe	16 Chief technician
17	Abteilung für Produktionsvorbereitung	17 Department for pre-production
18	Abteilung für Industrieökonomik	18 Department for budgeting
19	Arbeitsbüro und Lohnverrechnung	19 Labour office and pay roll accounting
20	Abteilung für Investition und Vermögensverwaltung	20 Department for investments and financial administration
21	Verwaltungsabteilung	21 Administration
22	Abteilung für Aufsicht und Revision	22 Department for supervision and auditing
23	Produktionsabteilung	23 Production department
24	Zentrale Qualitätskontrolle	24 Central quality control
25	Abteilung für Arbeitsschutz	25 Department for factory safety
26	Abteilung für Planung und Anschaffung	26 Department for planning and procurement
27	Abteilung für Mechanisierung der Geschäftsführung	27 Department for mechanization of management

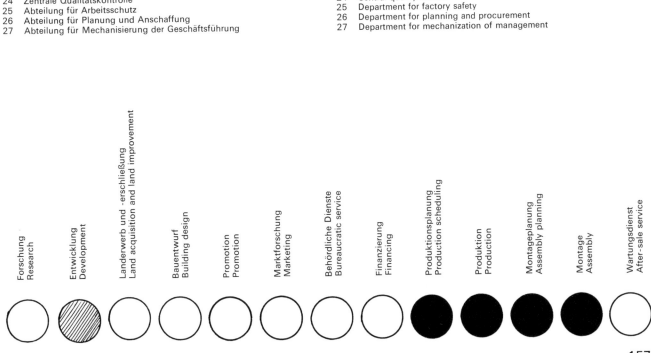

Forschung / Research
Entwicklung / Development
Landerwerb und -erschließung / Land acquisition and land improvement
Bauentwurf / Building design
Promotion / Promotion
Marktforschung / Marketing
Behördliche Dienste / Bureaucratic service
Finanzierung / Financing
Produktionsplanung / Production scheduling
Produktion / Production
Montageplanung / Assembly planning
Montage / Assembly
Wartungsdienst / After-sale service

5.3 Bemerkungen

Die Informationen in den Antworten der Firmen bieten meines Erachtens Stoff genug, um einige relevante Überlegungen anzustellen.

Gewiß variiert die Qualität der Information, ich erachte es aber als unnötig, auf einer konsistenten Qualität zu bestehen. Wir werden noch sehen, daß der Mangel an Informationen oder daß die oberflächliche Beantwortung gewisser Fragen trotzdem sehr aufschlußreich sein kann.

Bevor wir aber mit der Analyse dieser Daten beginnen, müssen wir einige Regeln festlegen und auf gewisse Gefahren hinweisen. Die erste und eindeutig erkennbare Gefahr, der wir ausgesetzt sind, besteht im Versuch, die verschiedenen Währungen miteinander vergleichen zu wollen. Jede Firma hat selbstverständlich ihre Informationen in der eigenen Währung wiedergegeben. Man könnte nun ohne weiteres diese Zahlen in eine einheitliche Währung umsetzen, was aber zu irreführenden Resultaten führen würde. Umrechnungskurse sind bekanntlich weit davon entfernt, ein genaues Bild von der relativen Kaufkraft einer Währung im eigenen Land zu verleihen. Noch schlimmer ist die Tatsache, daß die Kostenbelastung eines «typischen» Gebäudes im Jahresbudget der Käufer große Schwankungen aufweist. Überhaupt ist das «typische» Gebäude ein Mythos. Eine Familie der Mittelklasse wohnt in den USA meistens in einem Einfamilienhaus (von technisch niederem Niveau), das mit allen möglichen Einrichtungen ausgestattet ist (Klimaanlage, vollständig eingerichtete Küche usw.). Entsprechend wohnt eine Familie in Italien in einem mehrgeschossigen Wohnblock (dessen technisches Niveau viel höher liegt und der besser ausgeführt, aber lediglich mit elektrischen und sanitären Installationen versehen ist). Wiederum wird die entsprechende Familie in England in einem Reihenhaus wohnen.

Verfolgen wir diese Aspekte weiter, so kommen wir zum Schluß, daß wir völlig verschiedene Situationen miteinander vergleichen. Nehmen wir zum Beispiel die Kosten eines «Mobile Home» in den USA. Diese Behausung wird ungefähr auf 10 000 Dollar zu stehen kommen und enthält eine volleingerichtete Wohnfläche von ungefähr 80 m². Die Kosten dieses Hauses entsprechen dem Jahresgehalt des Käufers. Ein Käufer in Italien kauft dieselbe Wohnfläche in einem Wohnblock (und zwar ohne ein einziges Möbelstück) für etwa 15 000 Dollar. Dieser Betrag jedoch entspricht ungefähr einem Dreijahresgehalt.

Für unsere Zwecke ist eine derartige Information auch nicht direkt relevant, wir können aber einige Indexzahlen einführen, die den Vorteil haben, daß man die Resultate besser miteinander vergleichen kann.

Unter anderem besteht eines der Ziele unserer Untersuchung darin, die Leistungsfähigkeit der verschiedenen Organisationsformen aufzudecken und wenn möglich eine Art Muster oder Richtlinie zu extrapolieren.

Mit diesem Ziel im Auge können wir drei Indexziffern betrachten:

a) Investiertes Kapital – Wert der Produktion pro Jahr
b) Kosten der Arbeitskräfte – Wert der Produktion pro Jahr
c) Gebaute Quadratmeter pro Jahr – Zahl der Arbeitskräfte.

5.3 Remarks

I feel that the information made available by the analysed organizations offers sufficient material for a few relevant considerations.

Certainly the quality of the information varies, but I have not felt it necessary to insist on one consistant quality. As we shall see, the lack of some information or the superficial answer to some questions are, in themselves, useful.

However, before beginning our analysis we must set certain rules, and consider certain dangers. The first and most obvious trap we have to avoid is that of making any attempt to compare monetary figures. Logically enough each organization has given its information in its own currency. We could easily translate these figures into one basic currency, but only with misleading results. We all know that the rates of exchange are far from being a true reflection of the relative buying-power of a given currency in its own country. But even worse is the fact that the incidence of a "typical" building on the yearly budget of the buyer will vary considerably. Moreover the "typical" building is a myth. In the U.S.A. the standard middle-class family will live in a one-family house (of very poor technical standard) fitted with all kind of appliances (air-conditioning, fully-furnished kitchen, etc.). The corresponding family in Italy will dwell in a multi-storey apartment-house (of considerably higher technical quality, and better-finished) with no appliances but an electrical and plumbing network. Again the same family in the United Kingdom will live in a semi-detached house.

If we seriously analyse the matter we come to the conclusion that we are comparing different kinds of vegetables all together. As an example let us consider the cost of a standard mobile home in the U.S.A. This building will cost roughly ten thousand dollars and will provide some eighty square metres of fully-furnished dwelling. The cost of the dwelling is equivalent to one year of the buyer's salary. Now, the same kind of buyer in Italy will buy the same surface in an apartment house (with not a stick of furniture) for about fifteen thousand dollars. But this amount represents some three years of salary.

For our purposes we don't need anyhow this kind of information, but we can operate with a few indexes, which have the advantage of being partially self-correcting.

One of the objectives of our analysis is to discover the efficiency of the different organizations and if possible to extrapolate some kind of pattern or guide-line.

With this objective in mind we can consider three indexes:

a) Capital investment – production value per year
b) Manpower cost – production value per year
c) Square metres of building per year – number of employees.

Name der Firma / Name of company	Investition/Produktion Investment/Production		Arbeitskräfte/Produktion Manpower/Production		m²/Angestellter sq.m/Employee	
National Home	37/85	0,43	37/85	0,43	2 800 000/4 500	630 (2)
Delta Homes	3/114	0,026	17/113	0,15	9 700/150	62
IPI	5/10	0,50	25/100	0,25	100 000/750	130
MBM	2/6,5	0,31	15/65	0,23	70 000/400	175
Balency & Schuhl	36/180	0,20	6/18	0,33	230 000/2 200	100
Constructions Modulaires	1,4/73	0,02	156/7 300	0,02	120 000/66	1 800
Carl Möller	30/20	1,5	65/200	0,32	30 000/430	70 (3)
Nachbarschulte	18/100	0,18	13/100	0,13	15 000/128	116
IGECO	30/50	0,60	115/500	0,23	(1)	—
Vic Hallam	28/120	0,23	277/1 232	0,22	325 000/2 265	144
Servicio Técnico	4,5/100	0,04	7/100	0,07	20 000/19	1 050
Kombinat Nr. 3	15/70	0,21	283/700	0,41	500 000/4 000	125
Staatl. Bauges. Nr. 43	9248/19 922	0,47	2227/19 922	0,11	350 000/1 741	201

(1) Auf Grund unterschiedlicher Tätigkeiten nicht berechenbar.
(2) Die Lizenznehmer sind nicht mit eingerechnet.
(3) Auf Grund unterschiedlicher Tätigkeit mit Vorsicht zu betrachten.

(1) Not computable because of mixed activity.
(2) Franchised builders not included.
(3) To be taken with caution because of mixed activity.

Aus diesen drei Indexziffern ergeben sich folgende Resultate:

Die Auswertung der oben angeführten Daten ist aufschlußreich, aber mit Vorsicht zu betrachten.
Wir können die Tatsache wohl kaum ohne Einschränkung annehmen, daß die Arbeitskräfte der Firma Constructions Modulaires 1800 m² pro Jahr «produzierten» im Vergleich zur niedrigen Zahl von 175 m² pro Jahr von MBM.
Tatsächlich sind Constructions Modulaires (1800 m² pro Kopf) und Servicio Técnico (1050 m² pro Kopf) in einer besonderen Kategorie. Sie stellen jene Art von Firmen dar, die nicht produzieren, sondern sich auf die Arbeitskraft fremder Firmen und Bauunternehmer verlassen, selber aber sozusagen als «Gehirn» tätig sind. Warum kommt dann Constructions Modulaires in dieser besonderen Kategorie soviel höher hinaus als Servicio Técnico? Es sind zwei Faktoren, die diese Tatsache erklären: Constructions Modulaires besitzt einen Computer, der die Arbeit von etwa zwanzig Leuten ersetzt; ferner ist Constructions Modulaires eine ältere Firma. Der anfängliche Personalüberschuß, der für die technische Entwicklung und den geschäftlichen Start notwendig ist, ist bereits absorbiert.
Die Zahlen der anderen Firmen sind schon eher vergleichbar. Gewiß zeigen einige Firmen eine höhere Produktion pro Kopf auf, wiederum zeigt aber eine Analyse der ganzen Daten, daß der niedrigste Stand der Produktion mit einem breiteren Tätigkeitsfeld zusammengeht. Dennoch bleibt eine bemerkenswerte Ausnahme übrig: Delta Homes. Diese Firma produziert in der eigenen Fabrik ein Fertigprodukt, aber mit einem äußerst niedrigen Verhältnis von Investition/ Produktion: 0,026. Gehen wir aber diesem Verhältnis weiter nach, so bemerken wir, daß diese Firma in gepachteten Anlagen tätig ist, wodurch die Firma hohe Investitionen umgehen kann. Viele Teile des Gesamtproduktes werden auch von fremden Herstellern gekauft (z.B. die Stahlrahmen und einiges Zubehör). Es würde sich für die Konkurrenz dieser Firma lohnen, ein solches Vorgehen genauer anzusehen.
Ebenso aufschlußreich sind die fehlenden Angaben der Firma SEAC, der einzigen nicht auf Gewinn ausgerichteten Firma in unserer Untersuchung. Die anderen Indexzahlen bestätigen nur unsere eingangs gemachte Aussage, daß mögliche Fehler oder irreführ-

These three indexes give us the following results:

A review of the above data is interesting but again caution is necessary.
Clearly we cannot accept without qualifications the fact that the people of Constructions Modulaires "produce" 1800 sq.m per year against the poor 175 sq.m per year of MBM!
In fact, Constructions Modulaires (1800 sq.m per head) and Servicio Técnico (1050 sq.m per head) are in a league of their own. They represent the kind of companies which do not produce, but rely on outside manufacturers and building contractors for the physical activity, while they provide "the brain". Why then does Constructions Modulaires in this specific league score so much better than Servicio Técnico? Two factors explain this fact: Constructions Modulaires owns a computer which provides something like the equivalent of twenty people's work and is an older company. The initial overstaffing necessary for the technical development and commercial launching has already been absorbed.
In the other group of companies the results are more or less comparable. Certainly some companies show a higher productivity per head, but again an analysis of the complete data will show that the lowest productivity also coincides with a more varied range of services. A notable exception still remains: Delta Homes. This company produces in its own factory a finished product, but with an extremely low investment/production ratio: 0.026. A close analysis explains the situation. The company operates in leased premises, thus managing to avoid the high investment required by fully-owned facilities. Also a considerable amount of the product is bought from outside suppliers (i.e.: the steel frame and some fittings). The approach is definitely worth deeper analysis by some of the firm's competitors.
Also interesting is the non-availability of information from SEAC, the only non-profit organization included in our review. This case is, however, so important that it merits a separate analysis. The other indexes we have established only reconfirm our previous statement, ensuring that possible errors or misleading information will not alter the general picture. We can therefore conclude with some general statements:

rende Informationen unser allgemeines Bild nicht zu verfälschen vermögen. Wir können deshalb mit folgenden Bemerkungen schließen:
a) Nichtherstellende Firmen bewähren sich anscheinend am besten
b) Firmen, die «leichte» Baustoffe verarbeiten (Holz, Kunststoff), weisen eine höhere Produktivität pro Kopf und niedrigere Investitionsziffern auf als Firmen, die schwierige Produkte herstellen (Beton)
c) Pachten von Einrichtungen und von Behausungen verdienen ernsthafte Beachtung
d) Gemischte Betriebe (konventionell und industrialisiert) sind nicht unbedingt empfehlenswert.
Können wir nun von diesen allgemeinen Aussagen einige Richtlinien ableiten? Um diese Frage beantworten zu können, müssen wir auf einige theoretische Grundkonzeptionen zurückgreifen.
Experten haben versucht, die Vorteile der «horizontalen» Strukturorganisation mit denen der «vertikalen» zu vergleichen. In der horizontalen Struktur sind diverse unabhängige Firmen im Spiel. Einige Hersteller sorgen für Bauteile und/oder Subsysteme. Die Designs können durch unabhängige Büros besorgt werden. Errichtung und Montage werden anderen Spezialfirmen zugeteilt. Man könnte eine horizontale Struktur zur konventionellen Bauindustrie zählen, mit dem wichtigen Unterschied, daß hier Fertigkomponenten und nicht unbestimmte Baustoffe gebraucht werden (für korrekte Definition siehe Verzeichnis der Begriffe).
Im konventionellen Bau werden Backsteine oder Stahlteile gekauft, im industrialisierten Bau Wände oder Stahlstrukturen.
In einer vertikal strukturierten Firma sorgt diese allein für Grundelemente, Werkproduktion von Subsystemen und schließlich Montage auf der Baustelle und Errichtung der Subsysteme.
Das eindeutigste Beispiel einer horizontalen Firma ist in unserer Aufstellung Constructions Modulaires und das eindeutigste einer vertikalen Firma National Home.
Was trägt nun unsere Analyse zu diesem Thema bei? Um diese Frage zu beantworten, müssen wir auf Abschnitt 5.2 zurückgreifen. Es scheint logisch, daß eine horizontale Firma im Bereich der Produktion beträchtliche Vorteile aufweist. Jeder Hersteller spezialisiert sich in der Produktion auf ein gewisses Subsystem, wofür er seine Ausrüstung optimal ausnützen kann. Er kann weiterhin sein Produkt einem offenen Markt anbieten und so seinem Produktionspotential vollkommen gerecht werden.
Firmen, die sich der Baustellenmontage von Subsystemen widmen, ist es möglich, ihre Dienste einem weiten Kundenkreis anzubieten, und sie können auf diese Weise ebenfalls ihr Potential vollkommen ausschöpfen. Da die Firma, die das ganze Produkt (Haus, Schule usw.) fördert, nicht durch Einrichtungs-Investitionen gehemmt ist, kann sie die für eine bestimmte Aufgabe qualifiziertesten Hersteller und Monteure aussuchen.
Nach diesen Überlegungen drängt sich nun folgendes Modell auf:
a) Hersteller von Subsystemen (spezialisiert und in offener Konkurrenz)
b) «Promoting»-Firma (Wahl von Herstellern, Integration der Produkte, gesamthafte technische Aufsicht)
c) Monteure (auf der Baustelle unter Aufsicht der «Promoting»-Firma).

a) Non-manufacturing organizations seem to be the most valid kind of organization
b) Organizations operating with "easy" materials (timber, plastic) have higher productivity per head and lower investment-rates than organizations which manipulate "difficult" products (concrete)
c) Leasing of equipment and facilities is well worth serious consideration
d) Mixed operations (conventional and industrialized) are not necessarily a good approach.
Can we derive some basic guidelines from these general statements? To answer this question we have to fall back on some theoretical conceptions.
Experts have been attempting to compare the advantages of the "horizontal" structure versus the "vertical" structure. A horizontal structure is that in which the basic activities are performed by independent organizations. Some manufacturers will provide components and/or sub-systems. Designs may be provided by independent offices. Erection and assembly will be the responsibility of specialized companies. In a way a horizontal structure is the one existing in the conventional building industry, with the meaningful difference that finished components are provided rather than undetermined materials (see glossary for the correct definitions).
Therefore while in conventional building we may buy bricks or steel sections, in industrialized building we shall buy walls or steel structures.
In a vertical structure we find within a single organization the agencies providing basic components, the factory-production of sub-systems and finally the site-assembly and erection of the sub-systems. In our review the closest example of a horizontal organization is Constructions Modulaires, while the closest example of a vertical organization is National Home.
What does our analysis add to this subject?
To answer we must look into the answers given in 5.2. It seems logical that a horizontal organization offers considerable advantages in the production area. Each manufacturer specializes in the production of a certain sub-system. He can optimize his equipment for a certain kind of production. He can also offer his product to an open market, thus exploiting to the full his production potential.
The organizations devoted to the site-assembly of the sub-system are free to offer their services to a wide range of customers, achieving again a full utilization of their capacities. The pivot-organization, the one which promotes the complete product (house, school, etc.), being free of heavy investment burdens, can choose from the manufacturers the most suitable, and from the assemblers the most qualified for the specific job.
According to this line of reasoning we may recognize as desirable the following model:
a) Manufactures of sub-systems (specialized and in open competition)
b) Promoting organizations (selecting manufacturers, integrating the products, providing overall technical control)
c) Assemblers (operating on site, under the control of the promoting organization).
At this point, however, we have to accept that a horizontal organization is unsatisfactory in the "marketing" area. The market will always be outside its

Wir müssen aber die Tatsache hinnehmen, daß sich eine horizontale Firma im Bereich des Marketings als unbefriedigend erweist. Der «Markt» wird immer außerhalb ihrer Kontrolle stehen. Die lukrative Funktion der Promotion (wie in 5.1 definiert) und der Finanzierung liegt in fremden Händen. Auch sind Entwürfe, die von fremder Hand verfaßt werden, selten zufriedenstellend. Es scheint, daß eine vertikale Firma bessere Resultate erzielen sollte. Die «Promoting»-Firma, wie sie oben definiert wurde, sollte dann folgende Bereiche umfassen:
a) Entwicklung
b) Promotion
c) Marketing
d) Bauentwurf
e) Bürokratische Dienste
f) Planung der Montage
g) Finanzierung
h) Wartungsdienste

Stellen wir diese beiden Schlußfolgerungen graphisch dar, so erhalten wir folgendes Diagramm:

control. The lucrative function of promotion (as defined in 5.1) and of financing will be in foreign hands. Designs made by outsiders can hardly be satisfactory. It seems logical that a vertical organization should give better results. The "promoting" organization defined above should then cover:
a) Development
b) Promotion
c) Marketing
d) Building design
e) Bureaucratic service
f) Assembly planning
g) Financing
h) After-sale service.

If we bring graphically together the two conclusions we obtain the following diagram:

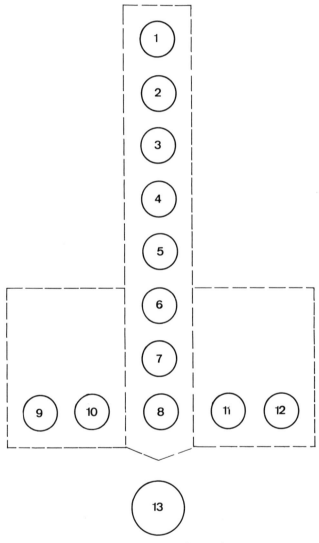

1 Entwicklung
2 Promotion
3 Marktforschung
4 Bauentwurf
5 Bürokratische Dienstleistungen
6 Montageplanung
7 Finanzierung
8 Wartungsdienst
9 Landerwerb
10 Herstellung der Komponenten
11 Landerschließung
12 Montage
13 Benutzer

1 Development
2 Promotion
3 Marketing
4 Building design
5 Bureaucratic service
6 Assembly planning
7 Financing
8 After-sales service
9 Land acquisition
10 Production of components
11 Land improvement
12 Assembly
13 User

Eine vertikal-horizontale Struktur scheint demnach optimal den Anforderungen zu entsprechen. Einige Firmen bewegen sich in dieser Richtung; andere neigen dazu, sich auf Anlagen und Technologie zu konzentrieren.
Vom gestalterischen Standpunkt aus ziehe ich eine möglichst weite und freie Wahl vor, und als Manager sehe ich darin keinen Vorteil, von einer teuren und nicht optimal ausgenützten Ausrüstung belastet zu sein.
Aufschlußreiche Informationen werden auch durch die Analyse der Antworten der Firma SEAC gewonnen. Diese Firma ist eines der mächtigsten britischen Konsortien, die sich der Aufgabe widmen, das Bedürfnis

We see then that a vertical-horizontal structure seems to satisfy optimally the requirements. Some organizations are advancing on this road; others tend to defend a more "material-based" position.
From a design point of view I favour the idea of a wide freedom of choice and as manager I do not see the advantage of being cluttered by expensive and under-utilized equipment.
Some interesting information may be obtained through the analysis of the answers given by SEAC. This organization is one of the powerful British consortia dedicated to satisfying the need for educational facilities. Basically all consortia operate on the same pattern, so that a generalization is acceptable. Besides a claim

nach Bildungsstätten zu stillen. Grundsätzlich arbeiten alle Konsortien nach demselben Muster, so daß sich eine Verallgemeinerung rechtfertigt. Neben dem Anspruch auf Qualität haben die Konsortien schon lange von sich behauptet, daß ihre Bauten billiger seien als solche, die auf dem offenen Markt gekauft werden. Diese Aussagen, verbunden mit der Tatsache, daß die Konsortien Großbritannien mit einer ausreichenden Anzahl von Bildungsstätten versehen, haben die Haltung der meisten Experten stark beeinflußt (auch mich). Durch meine direkte Erfahrung mit CLASP zweifle ich aber allmählich an der technischen Gültigkeit dieses Vorgehens. Wie bei den meisten Firmen sind die Konsortien durch überdurchschnittliche Persönlichkeiten gegründet worden, um spezifische Probleme unter ganz bestimmten Umständen anzupacken. Im Laufe der Zeit haben diese Firmen den üblichen Weg eingeschlagen: Sie haben die fähigsten Köpfe verloren und sind äußerst konservativ geworden. Technisch gesehen, sieht das Ergebnis so aus: Geringe Fortschritte werden gemacht, und neue Konzeptionen stoßen auf Widerstand. Dennoch wird hervorgehoben, daß die Bauten des Konsortiums billiger seien als andere. Was geschieht aber nun? Wenn wir die Antworten in 5.2 analysieren, dann sehen wir, daß SEAC die Kosten nicht zu berechnen vermag, die es braucht, um das «System» auf dem modernen Stand zu halten und es zu entwickeln (das investierte Kapital). Sie weiß auch nicht, was es kostet, die Firma zu leiten. Dieser Mangel an Informationen ist tatsächlich unvermeidlich. Die Arbeit wird von Beamten verrichtet, die ihre Kompensation durch die entsprechende Ortsbehörde erhalten und einen Teil ihrer Zeit für die Leitung der SEAC aufwenden. Wir *müssen* aber annehmen, daß die entsprechende Ortsbehörde ihr Personal reduzieren könnte, wenn nicht einige Beamte von Arbeiten für die SEAC beansprucht wären. Wir entdecken also, daß zu den offiziellen Baukosten noch versteckte Kosten (Leitung, technische Entwicklung) hinzukommen. Wären dann diese «wirklichen» Kosten immer noch niedriger als jene der privaten Firmen? In diesem Stadium ist die Frage schwierig zu beantworten; der Anspruch der Konsortien auf Wirtschaftlichkeit verliert aber dadurch gewiß etwas an Überzeugung. (Diese Ausführungen bedeuten nun keineswegs, daß SEAC und andere Konsortien Fehlschläge sind! Wir müssen im Gegenteil zugeben, daß die bedeutendsten Beiträge zur Industrialisierung des Bauens in den letzten fünfundzwanzig Jahren von den Konsortien stammen. Was wir aber gesagt haben, erhärtet eine bereits intuitiv erkannte Tatsache: Der Beitrag der Konsortien liegt nicht so sehr in den technischen oder wirtschaftlichen Errungenschaften ihres Produktes, sondern in der bloßen Tatsache, daß sie existieren und ein Beispiel für Vorgehensmethoden geliefert haben.)
Wer den organisatorischen Aufbau erforscht, für den kann die Analyse der Firmen in Kapital 5.2 von einigem Interesse sein. Dennoch zeigen jene, die korrekt geantwortet haben, die Wichtigkeit der kontrollierenden Funktion im Gesamtbild. Unser Anspruch in Kapitel 4, daß Industrialisierung hauptsächlich ein Problem des organisatorischen Aufbaus ist, scheint gut fundiert zu sein. Der Fachmann oder Techniker, der gewillt ist, neue technische oder technologische Konzeptionen zu erproben, sollte sich diesen Punkt wohl merken.

on quality the consortia have claimed throughout the years that their buildings are less expensive than similar buildings bought in the open market. These statements and the undisputable fact that the consortia have provided the United Kingdom with a satisfactory quantity of educational buildings have strongly conditioned the attitude of most experts (myself included). However, during my direct experience with CLASP I have started to doubt the general validity of the approach. As with most organizations the consortia have been created by above-average individuals to answer a specific challenge under a set of given circumstances. With the passing of time these organizations have followed the usual patterns: they have lost the most capable people and have become extremely conservative. The result, technically, has become evident: little progress has been made and new conceptions are hardly accepted. Still the point was made that the buildings produced by the consortia were less expensive than others. But now what happens? When we analyse the answers given in 5.2 we see that the organization does not know the cost of maintaining the "system" up to date, and of developing it (the capital investment). It also cannot indicate the cost of managing the organization. This lack of information is indeed unavoidable. The work is performed by public officials, who receive their compensation by the respective local authority, and who devote part of their time to the management of the SEAC organization. However, we *have* to assume that the respective local authorities could reduce their staff if some of their officials were not devoted to SEAC. We discover then that on top of the official cost of the building we have a hidden cost (management, technical development) which has to be added. Would this "real" cost still be lower than the one of the private organizations? The question at this stage is difficult to answer, but certainly the claim of cost-saving advanced by the consortia loses some of its strength.
(The above does not have to mean that SEAC and the other Consortia are failures! On the contrary, we have to admit that the most significant contributions to industrialized buildings in the last twenty-five years come from the Consortia. However, the information provided by SEAC strengthens an already existing opinion: the contribution of the Consortia lies not as much in the technical or economical quality of their product but in the mere fact of having existed and established a pattern of performance.)
To the student of organizational set-ups the review of the organizations in 5.2 can be of some interest. The ones who have provided complete information indicate how important the controlling function is in the general picture. I feel that our claim in chapter 4, that industrialization is mainly a problem of organization, becomes well substantiated. The practitioner or the technician who is willing to gamble purely on the strength of a new technical or technological conception should well consider this point.

6. Meinungen der Experten

6. Opinions of Experts

6.1 Einleitung

Ich hielt es für nützlich, in diesem ziemlich theoretischen Werk einige Meinungen von anerkannten Fachexperten hinzuzufügen. Hierzu habe ich die wichtigsten Fragen an meine Kollegen gerichtet, ohne aber zu versuchen, die Antworten in Einklang zu bringen. In einem Gebiet wie der Industrialisierung des Bauens, das sich immer noch in der Entwicklung befindet, scheint es nur natürlich zu sein, daß die Meinungen auseinandergehen und sich sogar direkt widersprechen können. Auch die Kenntnis solcher Divergenzen kann sich als fruchtbar erweisen.
Ich habe auf eine Zusammenfassung der vorgetragenen Meinungen am Ende dieses Kapitels verzichtet.
Erstens wäre es nicht angebracht, die Aussagen zu kritisieren, ohne die Experten replizieren zu lassen. Zweitens wäre eine Zusammenfassung verfehlt, da alle Äußerungen innerhalb eines gegebenen Kontextes gültig sind.
Alle Äußerungen stammen aus der Zeit zwischen Juli 1971 und Januar 1972.

6.1 Introduction

I have considered it useful to add to this rather theoretical work, the opinions of recognised experts in the field. I have therefore selected a few of the most obvious questions and I have asked my colleagues to answer them without attempting to reach a common agreement. In a field like the industrialization of building which still is in an evolutionary stage it seems logical, and useful to know, that opinions may differ and at times be even in strong opposition.
There is no summary of the given opinions at the end of this chapter.
First of all it would be unfair to criticize the opinions expressed by the experts without giving them a chance to reply. Secondarily a summing-up does not make much sense: all opinions are valid within a given context.
All opinions have been given in the period July 1971 – January 1972.

Anthony Williams

England, Architekt, ist Partner in einer Firma, die sich auf Erforschung, Gestaltung, Entwicklung und Dokumentation von Komponenten, Systemen, Techniken und Ausrüstung des Bauens spezialisiert.

1. Definieren Sie «die Industrialisierung des Bauens»
«Die industrielle Organisation des Bauens – durch Anwendung der besten Methoden und Techniken auf den integrierten Prozeß von Nachfrage, Gestaltung, Herstellung und Konstruktion.» RIBA, «The Industrialization of Building», 1965.

2. Welches sind die erfolgreichsten Unternehmen des industrialisierten Bauens?
Der Schulenbau in Großbritannien seit 1946, besonders aber das Hertfordshire-Schulprojekt, das sich zu einer Zeit entwickelt hat, als ein akuter Mangel an Baustoffen und Arbeitskräften auf den Bauplätzen herrschte, bevor das industrialisierte Bauen national und international anerkannt wurde.

England, architect, is a partner in a firm that specialises in the investigation, design, development, and documentation of components, systems, techniques and equipment for building.

1. Define "the Industrialization of Building"
"The organization of building industrially – by applying the best methods and techniques to the integrated process of demand, design, manufacture and construction."
RIBA, *The Industrialization of Building*, 1965.

2. Which have been the most successful industrialized building ventures?
U.K. Schools programme 1946 onwards, and particularly the early Hertfordshire Schools programme which was developed at a time when there was an acute shortage of building materials and site labour, before national and international appreciation of the need for industrialization awoke.

3. Nach welchen Maßstäben definieren Sie den Grad des Erfolges?
Insofern Bauteile in der Fabrik montiert werden können, mit der Absicht, den Zeitverlust auf der Baustelle bei schlechtem Wetter zu reduzieren und die Arbeitsbedingungen zu verbessern.

4. Wie unterscheiden sich industrialisierte Bauten von konventionellen Bauten?
—

5. Welche Vorteile bietet die Bauindustrialisierung im Vergleich zu heutigen konventionellen Methoden?
Bauelemente und Organisationsformen über längere Zeit entwickeln, prüfen und verbessern zu können.

6. Wie kann der Architektenberuf am besten einen Beitrag zum Industrialisierungsprozeß leisten?
Indem er sich dem Konzept der industrialisierten Methode und den daraus resultierenden Komponenten und Techniken nicht verschließt, sondern öffnet und sie wenn immer möglich anwendet, um die besten Ergebnisse zu erzielen.

7. Welche Rolle spielt die politische Macht – oder welche Rolle kann sie spielen –, um die Bauindustrialisierung zu fördern?
Eine beständige, kontinuierliche Nachfrage zu garantieren.

8. Welche Form der Bauindustrialisierung scheint Ihnen am meisten zu versprechen?
Alle Formen der Industrialisierung sind wertvoll und können je nach Umständen große Erfolge verzeichnen. Die Nachfrage schwankt aber beträchtlich. «Neue Techniken», die hoch geschätzt werden, wenn die Nachfrage am größten ist, fallen weg, wenn die Nachfrage nachläßt, zugunsten von traditionellen Methoden. Ich ziehe deshalb eine Konzentration auf die Entwicklung von Komponenten vor, welche bei der Vorfertigung, den Bausystemen, der rationalisierten traditionellen und sogar der traditionellen Bauweise angewendet werden können.

9. Bemerkungen
Schlagwörter wie Industrialisierung liegen mir eigentlich nicht. Sie werden gewöhnlich von Politikern derart aufgeblasen, bis sie – wie Ballone – bersten, so daß ein sachlicher Begriff in Verruf gerät. Wir müssen möglichst jede Technik ausnutzen, um die Qualität und Wirtschaftlichkeit des Bauens, besonders bei sozialen Bauten, zu verbessern und um den Bewohnern eine menschlichere Umwelt zu schaffen. Es besteht ein großes Bedürfnis nach besserem Management und vermehrter Industrialisierung.

3. By which parameters do you define the degree of success?
In the degree to which parts of buildings may be assembled within the factory, with the intention of reducing wastage of site time due to bad weather and of improving working conditions for manpower.

4. In what ways do industrialized buildings differ from conventional buildings?
—

5. What can industrialized building offer that conventional methods cannot provide today?
The possibility of developing, testing and improving components and organizations over a period of time.

6. How can the architectural profession best contribute to the industrialization process?
By having sympathy with and accepting the concept of industrialized methods and the components and techniques which result and by designing with them in the most sympathetic way to achieve the very best results.

7. What is or can be the rôle of the political power concerned in order to promote the industrialization of building?
To promote continuity and consistency of demand.

8. Which form of industrialization do you consider the most promising?
The varying forms of industrialization all have their uses and can be highly successful according to circumstances. But demand fluctuates very considerably. "New techniques" which are highly appreciated when the demand is at its greatest are abandoned in favour of traditional methods when demand drops. I therefore favour a large scale approach towards component developments, which can be associated with prefabrication, system building, rationalized traditional building and even traditional building.

9. Remarks
I do not like such catch phrases as industrialization. They have a habit of becoming blown up by politicians until, like a balloon, they are burst and an objective concept comes into disrepute. We need to use every possible technique to improve the quality and economy of building particularly for social building, and to provide a more sensitive environment for people. There is a great need for improved management and for more industrialization.

Thomas Schmid

Schweiz, Architekt, Designer von Bausystemen.

1. Definieren Sie «die Industrialisierung des Bauens»
Planung und Produktion von Bauwerken mit dem Ziel, industriell gefertigte Bauteile einzusetzen.

Switzerland, architect, designer of building systems.

1. Define "the Industrialization of Building"
Planning and production of building projects with the aim of employing industrially fabricated components.

2. Welches sind die erfolgreichsten Unternehmen des industrialisierten Bauens?
Im Wohnungsbau sind für mich die bisher erfolgreichsten Anwendungen von industrialisierten Baumethoden die Holzrahmenpaneelsysteme, in der Art von National Home, USA, und einiger schwedischer Systeme. Im Schulenbau steht meines Erachtens das System CLASP/Brockhouse an der Spitze.

3. Nach welchen Maßstäben definieren Sie den Grad des Erfolges?
Ein System soll
a) technisch so ausgereift sein, daß es den bautechnischen Anforderungen des betreffenden Landes genügt;
b) die funktionellen Anforderungen des betreffenden Landes voll erfüllen;
c) in der Bauform der gestellten Aufgabe gerecht werden.

4. Wie unterscheiden sich industrialisierte Bauten von konventionellen Bauten?
Konventionelle Bauten werden von Architekten nach individueller Auffassung entworfen, industriell produzierte Bauten werden nach den Gesetzmäßigkeiten des verwendeten Systems konzipiert, ohne individuelle Interpretationsmöglichkeiten für den Architekten. Ebenso entfällt die herkömmliche Art der Bauorganisation und des Kostenaufbaus.

5. Welche Vorteile bietet die Bauindustrialisierung im Vergleich zu heutigen konventionellen Methoden?
Größeres Bautempo.
a) Bessere Kostenkontrolle, weil sich der Kostenaufbau aus Preislisten der Bauelemente und Montagekosten zusammensetzt, die transparenter sind als die bisherige Kostenstruktur des Baugewerbes.
b) Weniger Bauschäden, weil ein Bausystem in der Regel besser durchgetestet ist als die stets wechselnden herkömmlichen Baumethoden.

6. Wie kann der Architektenberuf am besten einen Beitrag zum Industrialisierungsprozeß leisten?
Bisher war der Beitrag der Architekten eher klein, weil die Architekturschulen die neuen Methoden nicht lehrten. Vorerst sollte einmal der Architekturunterricht verbessert und modernisiert werden.

7. Welche Rolle spielt die politische Macht – oder welche Rolle kann sie spielen –, um die Bauindustrialisierung zu fördern?
Der Einsatz politischer Macht ist entscheidend. In den osteuropäischen Ländern führte er zu einer vollständigen Industrialisierung des Bauwesens, während im Westen der Widerstand des herkömmlichen Baugewerbes den Fortschritt stark bremste. Solange die Produktionsmittel in privaten Händen liegen, wird der Übergang zu industrieller Produktion sehr langsam vor sich gehen.

8. Welche Form der Bauindustrialisierung scheint Ihnen am meisten zu versprechen?
Die Zielform soll bestimmt das Componenting sein, mit freier Austauschbarkeit der Bauteile. Diese Stufe setzt jedoch einen breiten Markt, eine hochentwickelte Fugentechnik und genügend geschulte Architekten

2. Which have been the most successful industrialized building ventures?
In the field of housing, in my opinion, the hitherto most successful application of industrialized building methods is the wood-frame panel systems, such as are employed by National Home, U.S.A., and a number of Swedish systems. In school construction, I put the CLASP/Brockhouse System on top.

3. By which parameters do you define the degree of success?
A system ought:
a) To be so technically advanced that it meets the construction requirements prevailing in the given country.
b) To satisfy completely the functional requirements of the given country.
c) To do justice to the aim set in the given type of building.

4. In what ways do industrialized buildings differ from conventional buildings?
Conventional buildings are designed by architects in accordance with individual notions, while industrialized buildings are conceived in accordance with the norms of the system applied, without any scope for individual interpretation by the architect. The same thing goes for the traditional form of building organization and cost structure.

5. What can industrialized building offer that conventional methods cannot provide today?
Increased speed of construction.
a) Better cost control, because the cost structure consists of price lists of building components and assembly costs, which are more transparent than the old-style cost structures of the building trade.
b) Less damage during construction, because a building system, as a rule, has undergone stricter testing than the constantly changing traditional methods.

6. How can the architectural profession best contribute to the industrialization process?
Up to now the architect's contribution has been rather modest, because the schools of architecture have not taught the new methods. First of all, architectural training should be improved and modernized.

7. What is or can be the rôle of the political power concerned in order to promote the industrialization of building?
Political intervention is decisive. In the Eastern European countries it has led to total industrialization of building, while in the West the resistance of the traditional building trade has been a strong brake on progress. As long as the means of production are in private hands, the conversion to industrialized production will go forward very slowly.

8. Which form of industrialization do you consider the most promising?
The form to be aimed at ought definitely to be componenting, with free interchangeability of parts. This stage, however, presupposes an extensive market, a highly developed jointing technique and sufficiently trained architects. In most of the Western European

voraus. In den meisten Ländern Westeuropas sind diese Bedingungen aber noch nicht erfüllt, so daß man vorerst einmal mit den geschlossenen Systemen beginnen sollte.

countries, these conditions are not yet satisfied, so that for the time being a start should be made with the closed systems.

Masayumi Yokoyama

Ingenieur, ist einer der «Anstifter» der Bauindustrialisierung in der französischen Schweiz. Er war an der Experimentierung mit Betonfertigteilen beteiligt (1958).

1. Definieren Sie «die Industrialisierung des Bauens»
Die Industrialisierung des Bauens ist die Zusammenwirkung der menschlichen und materiellen Mittel, die von Behörden und/oder Privatunternehmen zum Zwecke einer besseren, schnelleren und billigeren Bautätigkeit angewandt werden. Die Hierarchie der unmittelbaren Ziele der Bauindustrialisierung sowohl wie der beruflichen, sozialen und wirtschaftlichen Strukturen, innerhalb deren sie sich entwickelt, hängt in höchstem Maße von politischen Umständen ab, denn diese beeinflussen den Staat, die Unternehmen oder die Fachleute. Sie bedingen das Ausmaß der Industrialisierung und den Erfolg der zahlreichen Techniken, die jetzt schon zu ihrer Verfügung stehen. Sie sind die Erklärung für die großen Unterschiede, die heute zwischen einzelnen Ländern bestehen.

2. Welches sind die erfolgreichsten Unternehmen des industrialisierten Bauens?
Auf den ersten Blick sehe ich keines, das heute wirklich «erfolgreich» genannt werden könnte; eigentlich sehe ich nur Beispiele, die Teilerfolge oder halbe Fehlschläge sind.

3. Nach welchen Maßstäben definieren Sie den Grad des Erfolges?
Nach der bestmöglichen Erfüllung der Bedürfnisse der Gemeinschaft und des Einzelnen (Mieter, Schüler, Patient usw.).

4. Wie unterscheiden sich industrialisierte Bauten von konventionellen Bauten?
Durch statische und dynamische Wirksamkeit — das heißt in Raum (Form und Funktion) und Zeit (Dauerhaftigkeit oder Veränderlichkeit).

5. Welche Vorteile bietet die Bauindustrialisierung im Vergleich zu heutigen konventionellen Methoden?
An erster Stelle Reduktion der Arbeitskräfte, zunächst auf der Baustelle, dann im Gesamtprozeß. Ebenso Kosten; oft bessere Leistung.

6. Wie kann der Architektenberuf am besten einen Beitrag zum Industrialisierungsprozeß leisten?
In der Spezialisierung; es sollte nicht «den» Architekten, sondern «die» Architekten geben: Architekt-Stadtplaner, Architekt-Ökonom, Architekt-Organisator, Architekt der Produkte usw. ... und aus all diesen Architekten werden einige *Architekten* hervorgehen — und die werden hervorragend sein.

Engineer, is one of the organizers of the industrialization of building in French-speaking Switzerland. He was behind the first works for producing concrete prefab components (1958).

1. Define "the Industrialization of Building"
The industrialization of building is the cooperation of human and material means that are applied by public authorities and/or private enterprises for the purpose of getting better, faster and cheaper building done. The hierarchy of the immediate goals of building industrialization as well as of the professional, social and economic structures, within which it develops, depends largely on political circumstances, for the latter influence the state, the building enterprises or technical people. They condition the extent of industrialization and the success of the numerous techniques which are already available. They provide the explanation for the great differences existing at the present time among individual countries.

2. Which have been the most successful industrialized building ventures?
At first glance I do not see any that at present could really be called "successful"; actually, I see only projects that are partial successes or semi-failures.

3. By which parameters do you define the degree of success?
In terms of the optimum compliance with the needs of the community and the private individual (tenant, pupil, hospital patient, etc.).

4. In what ways do industrialized buildings differ from conventional buildings?
By static and dynamic effectiveness — i.e., in space (form and function) and time (durability or alterability).

5. What can industrialized building offer that conventional methods cannot provide today?
In the first place, reduction of labour requirements, on the building site, then in the total process. The same goes for costs; often enhanced performances.

6. How can the architectural profession best contribute to the industrialization process?
By way of specialization; there should no longer be "the" architect, but the *architects*: architect-townplanner, architect-economist, architect-organizer, product architect, etc. ... and a few *architects* will emerge from all these architects — and they will be outstanding.

7. What is or can be the rôle of the political power

concerned in order to promote the industrialization of building?
My experiences and my present knowledge do not permit me as yet to answer this question except within a very limited and very specific context that does not admit generalizations.

8. Which form of industrialization do you consider the most promising?
Systems and componenting, backed up by a very strong business organization.

9. Remarks
1. As in many other fields, in building what is always stressed is productivity; could not genuine building industrialization — which has yet to be realized — put the accent on "quality", and how could this be done?
2. Should the "concentration" on the specialized level (building enterprises, research bodies, and building and loan societies) develop vertically or horizontally?

Wolfgang Döring

German Federal Republic, architect. Designer of building systems. Author of numerous publications on industrialized building.

1. Define "the Industrialization of Building"
The organization of building for the shaping of material by means of capital-intensive methods such as mechanization and automatization in a cybernetically controlled process.

2. Which have been the most successful industrialized building ventures?
I do not know of any, since up to now industrialized building has been carried out on an almost exclusively quantitative basis, to the total disregard of the necessary qualitative aspects — in this way our environment is destroyed by building. Pollution by building!

3. By which parameters do you define the degree of success?
In the relation of quantity to standardization. In relation to socially relevant results: Standardization can (in this context) only show human results if it is understood as a cybernetic model, whereby the results must in every case influence the new sociological programmes and whereby results, as almost always happens, only alter the technological programmes via feed-back.

4. In what ways do industrialized buildings differ from conventional buildings?
When in this context I think of conventional buildings, there occur to me the concrete monuments of "Brutalism", which have very rapidly lost their significance

Wenn ich in diesem Zusammenhang an konventionelles Bauen denke, fallen mir die Betondenkmäler des Brutalismus ein, die sehr schnell ihre Bedeutung durch den optischen Verschleiß der Kommunikationsmedien eingebüßt haben. Industrialisiertes Bauen zeigt die Komplexität eines Hauses aus Installation, Konstruktion, Kommunikationsmechanik, modularer Ordnung, Montage, Demontage, Umsetzbarkeit, Reproduzierbarkeit, Einrichtung, Materialkontrolle, Elementierung, Schallkontrolle, Isolationskontrolle usw.

Ein industrialisiertes Haus ist ein Industrieprodukt, entstanden aus der Kenntnis und der vollständigen Beherrschung der Werkzeuge unserer technisch orientierten Zivilisation. Im Gegensatz zu konventionellem ist industrialisiertes Bauen Veränderungen gegenüber offen. So daß die so entstehenden Räume den sich ständig ändernden Forderungen flexibel angepaßt werden können.

5. Welche Vorteile bietet die Bauindustrialisierung im Vergleich zu heutigen konventionellen Methoden?
—

6. Wie kann der Architektenberuf am besten einen Beitrag zum Industrialisierungsprozeß leisten?
Diese beiden Fragen gehören für mich zusammen.
Nur der Architekt ist in der Lage, der Industrialisierung, der Standardisierung den notwendig qualitativen Aspekt zu geben, um Bauen in einer rein statistischen Dimension zu verhindern, um so zu einer Humanisierung industrieller Systeme zu gelangen.
Aufgabe des Architekten, des unabhängigen Planers, ist es, durch qualitative Veränderungen die Uniformität der Umwelt zu verhindern, Strukturen zu schaffen, welche neue Grenzen und Horizonte durch das Regulativ der Ästhetik sichtbar werden lassen. Das kann nur erreicht werden mit industriellen Methoden, den Werkzeugen unserer Zeit. Industrialisierung heißt Elementierung der einzelnen Teile des Gesamtkomplexes in der Weise, daß jeweils andere, voneinander unterscheidbare und damit identifizierbare, architektonisch relevante Räume machbar — konstruierbar sind.

7. Welche Rolle spielt die politische Macht — oder welche Rolle kann sie spielen —, um die Bauindustrialisierung zu fördern?
Der Politik fällt die Schlüsselrolle zu, denn hier wird die Strategie entwickelt für das Programm.
Das Problem der Grundlagenforschung: wie bei dem Flug zum Mond, der mit Steuergeldern finanziert wurde, so wird es auch hier unumgänglich sein, durch politische Einflußnahme die Öffentlichkeit an den programmatischen Voraussetzungen zu beteiligen. Notfalls durch eine Änderung der gesellschaftlichen Situation.

8. Welche Form der Bauindustrialisierung scheint Ihnen am meisten zu versprechen?
Es gibt hier keine besondere Art von Industrialisierung, die im Zusammenhang mit Bauen ein Umdenken voraussetzt, und zwar Bauen als Vorgang — und nicht als statische Dimension. Industrialisierung ist ein flexibles Werkzeug unter Ausnützung sämtlicher Möglichkeiten, um die Formen soziologischer Strukturen in eigener Verantwortung auszuwählen und bestimmen zu können.

owing to their being overworked by the mass media. Industrialized building shows the complexity of a house, made up of the following aspects: installation, construction, communications technique, modular organization, assembly, dismantling, convertibility, reproducibility, fitting, materials, components, acoustic factor, insulation factor, etc.

An industrialized house is an industrial product, arising out of the knowledge of and complete mastery of the tools of our technologically oriented civilization. In contrast to conventional building, industrialized building is open to alterations. The result is that volumes can be adapted to constantly changing requirements.

5. What can industrialized building offer that conventional methods cannot provide today?
—

6. How can the architectural profession best contribute to the industrialization process?
These two questions for me belong together.
Only the architect is in a position to give to industrialization, to standardization, the necessary qualitative aspect, in order to prevent the rise of an architecture that remains stuck in a purely statistical dimension, in order to humanize industrial systems.
The rôle of the architect, the independent planner, is by way of qualitative changes to prevent the rise of a uniform environment, to create structures which reveal new frontiers and horizons. That can be achieved only by industrial methods, the tools of our age. Industrialization means the norming of the individual parts of the total complex in such a way that always new, distinct and thus identifiable, volumes that are architecturally relevant can be constructed.

7. What is or can be the rôle of the political power concerned in order to promote the industrialization of building?
The state is in a key position, for here is where the strategy is developed for the programme.
The problem of fundamental research: as in the case of the moon shots, which have been financed by tax proceeds, here too it will be absolutely imperative for the public to exert its political influence on the programmes. If need be, by a change in the social structure.

8. Which form of industrialization do you consider the most promising?
There is no special form or type of industrialization which, in connection with building, calls for revision of our thinking, and I mean building as a process — and not as a static dimension.
Industrialization is a flexible tool employing all potentialities in order to select the forms of social structure on its own responsibility and to be able to determine them.

David Chasanow

UdSSR, Architekt, Leiter der Normierungsabteilung des Zentralinstitutes für Forschung im Wohnungsbau, Moskau.

1. Definieren Sie «die Industrialisierung des Bauens»
Industrialisierung ist die Anwendung von Methoden der mechanischen Produktion im Bereich des Bauens.

2. Welches sind die erfolgreichsten Unternehmen des industrialisierten Bauens?
Die optimale Ausnützung der Industrialisierung liegt zwischen zwei Extremen: vollständige Vorfertigung in der Werkhalle und industrialisiertes Bauen auf der Baustelle. Das Gleichgewicht verschiebt sich je nach der Entwicklung der Technik und Wissenschaft.

3. Nach welchen Maßstäben definieren Sie den Grad des Erfolges?
Der Erfolg wird mit den traditionellen Maßstäben der Architektur gemessen, die bereits zur Zeit des Vitruvius formuliert wurden: Nützlichkeit (Komfort), Dauerhaftigkeit, Schönheit und Wirtschaftlichkeit. Der Beginn der technischen Revolution im Baugewerbe stand im Zeichen des Vorranges von wirtschaftlichen Kriterien, während heute die Begriffe Quantität und Qualität unzertrennlich miteinander verbunden sind.

4. Wie unterscheiden sich industrialisierte Bauten von konventionellen Bauten?
Der Einfluß der modernen Industrie auf die Architektur manifestiert sich in der begrenzten Auswahl von Modulardimensionen, in der Vorliebe für einfache Formen, in der Notwendigkeit der teilweisen oder vollständigen Normierung. Der Geist der Architektur war schon immer vom Niveau des technischen Fortschrittes, von wirtschaftlichen Möglichkeiten und anderen Bedingungen abhängig. Man hat damals, als die Situation noch nicht so komplex war, Kunstwerke geschaffen.

5. Welche Vorteile bietet die Industrialisierung im Vergleich zu heutigen konventionellen Methoden?
Es waren die handwerklichen Produktionsgesetze, die die Ästhetik der traditionellen Bauweise bestimmten. Die maschinelle Produktion hat einige neue Elemente hinzugefügt — die Ästhetik der Oberflächen und perfekte Linienführung ersetzen den Charme der kapriziösen Formen der manuellen Arbeitsweise.

6. Wie kann der Architektenberuf am besten einen Beitrag zum Industrialisierungsprozeß leisten?
Das Konstruktionsziel ist, Bauwerke und Städte zu bauen. Deshalb muß der Architekt die Gesetze der Industrie kennen, auf ihre Gesetzmäßigkeit eingehen und seine Arbeit danach richten.

7. Welche Rolle spielt die politische Macht — oder welche Rolle kann sie spielen —, um die Bauindustrialisierung zu fördern?
Die Planung und die Koordination, die von den Staatsorganen ausgehen, sind gegenwärtig die notwendigen Bedingungen der Entwicklung der Architektur, die nur beschränkt mit allen Zweigen der Industrie, der

U.S.S.R., architect, head of the Division of Standardization of the Central Research Institute for Housing Design, Moscow.

1. Define "the Industrialization of Building"
Industrialization is the application of the methods of mechanical production to the field of building.

2. Which have been the most successful industrialized building ventures?
The optimum kind of industrialization is to be found between two extremes: total prefabrication at works and industrialized construction in situ. The path to be followed changes as technology and science change.

3. By which parameters do you define the degree of success?
Success ought to be measured in terms of the traditional criteria of architecture formulated as long ago as the time of Vitruvius: utility, solidity, beauty and economic reasons. The early period of the technical revolution in the field of building was characterized by the predominance of economic criteria, but at the present time the notions of quantity and of quality are inseparably linked.

4. In what ways do industrialized buildings differ from conventional buildings?
The influence of modern industry on architecture is shown in the clear grading of modular dimensions, the preference for simple shapes, the necessity for a certain degree of standardization. Nevertheless, the creative genius of the architect was at all times limited by the level of technological development, economic potentialities and other objective conditions; in this situation that was no less complex than that of nowadays, masterpieces were created.

5. What can industrialized building offer that conventional methods cannot provide today?
The laws of handicraft production determined the aesthetics of traditional architecture. Mechanical production introduced some new elements — the aesthetic principle embodied in flawlessly exact surfaces and lines replaced the charm of capricious shapes created by handicraftsmen.

6. How can the architectural profession best contribute to the industrialization process?
The aim of building is to create the building, the city, architecture. That is why the architect ought to be familiar with the laws governing industry, to comply with its exigencies and resume control of building projects.

7. What is or can be the rôle of the political power concerned in order to promote the industrialization of building?
Planning and coordination carried out by state organs are at the present time the necessary conditions for the development of architecture, which is closely bound up with all branches of industry, science, engineering, art.

Wissenschaft, der Technik und der Kunst verbunden ist.

8. Welche Form der Bauindustrialisierung scheint Ihnen am meisten zu versprechen?
Die rationellen Formen der Industrialisierung bleiben nicht ewig. Gegenwärtig ist es die vorgefertigte Bauweise mit großen Paneelen oder mit Zellen aus Stahlbeton, morgen vielleicht vorgefertigte Wohneinheiten, die in ein tragendes Skelett eingehängt werden, und übermorgen vielleicht wieder etwas anderes. Leben und Denken stehen nie still.

8. Which form of industrialization do you consider the most promising?
The rational forms of industrialization are not eternal. Today we have prefab construction of large slabs or boxes of reinforced concrete; tomorrow perhaps we shall see prefab houses suspended from some kind of carrying structure, the day after tomorrow something else again. Life and thought never come to rest.

Jean Prouvé

Frankreich, Architekt und Unternehmer, Paris

France, architect and enterpriser, Paris.

1. Definieren Sie «die Industrialisierung des Bauens»
Wie kann man eine gebieterische Notwendigkeit definieren! Es ist eindeutig, daß nur eine in der Entwicklung begriffene Industrie «den Dienst leisten» kann, auf dynamische Weise der Menschheit die Behausung zu schaffen, die der immer mehr anwachsenden allgemeinen Evolution angepaßt ist.

1. Define "the Industrialization of Building"
How can one define an imperious necessity! It is perfectly clear that only one developing industry can *"perform the service"* of creating for the human race in a dynamic way the housing which is adapted to our ever accelerating evolution.

2. Welches sind die erfolgreichsten Unternehmen des industrialisierten Bauens?
Kann man von Erfolgen sprechen?
Ist es eine zu scharfe Kritik zu sagen, daß die Realisierungen bis zum heutigen Tag noch eine Verknüpfung mit der Vergangenheit verraten, die mir unvereinbar mit der Produktion im industriellen Rhythmus zu sein scheint. Gewiß, es gibt Schulbauten wie zum Beispiel in Frankreich, die sich weder von den halbindustrialisierten noch von den konventionellen Gebäuden unterscheiden.

2. Which have been the most successful industrialized building ventures?
Can one speak of successes?
Is it being too sharply critical to say that realizations to date still betray a tie with the past which strikes me as incompatible with production at an industrial pace. To be sure, there are school buildings, e.g., in France, which are not different either from semi-industrialized or from conventional buildings.

3. Nach welchen Maßstäben definieren Sie den Grad des Erfolges?
Durch die Finanzierung angewandter Forschung, realisiert im Rahmen einer Umgruppierung der Kompetenzen, die vollkommen in eine gemeinsam geschaffene Industrie integriert sind. Genauso wie sich wohlbekannte industrielle Wunder verwirklicht haben.

3. By which parameters do you define the degree of success?
By the financing of applied research carried out within the scope of a reorganization of competences which are perfectly integrated in a jointly created industry. Just as famous industrialists have worked well-known miracles.

4. Wie unterscheiden sich industrialisierte Bauten von konventionellen Bauten?
Bis jetzt, abgesehen von einigen noch nicht industrialisierten Modellen, unterscheiden sie sich sehr wenig, es sei denn durch eine ziemlich allgemeine Abwesenheit der Sensibilität.

4. In what ways do industrialized buildings differ from conventional buildings?
Up to now, except for a few not yet industrialized models, they differ very little, aside from a rather general absence of sensibility.

5. Welche Vorteile bietet die Bauindustrialisierung im Vergleich zu heutigen konventionellen Methoden?
Sie müßte folgendes anzubieten haben:
a) Mobilität durch Leichtigkeit
b) Möglichkeit der Landwiedergewinnung
c) Architektur, die ihrem Zeitalter ehrlich entspricht
d) Schnelle Amortisation.

5. What can industrialized building offer that conventional methods cannot provide today?
It would have to offer the following:
a) Mobility owing to lightness
b) Possibility of reacquiring land area
c) Architecture that honestly corresponds to its age
d) Rapid amortization.

6. Wie kann der Architektenberuf am besten einen Beitrag zum Industrialisierungsprozeß leisten?

6. How can the architectural profession best contribute to the industrialization process?
By complete integration in industry, i.e., production plants.

7. What is or can be the rôle of the political power

Durch vollkommene Integration in die Industrie, das heißt die Fabrik.

7. Welche Rolle spielt die politische Macht — oder welche Rolle kann sie spielen —, um die Bauindustrialisierung zu fördern?
Die Verstädterung ist Angelegenheit der Politik: Behausung und ihre Erweiterung gehen aus ihr hervor. Kann man eines Tages auf einen öffentlichen Dienst hoffen, für den das Recht zur Behausung eine Pflicht wird? In dem Falle hätte die Industrialisierung ihren Markt gefunden.

8. Welche Form der Bauindustrialisierung scheint Ihnen am meisten zu versprechen?
Alle industrialisierten Produktionen von großem Ausmaße sind sehr vielseitig; daraus kann man schließen, daß weder das System noch die Elemente noch die Techniken das Vorrecht haben.

9. Bemerkungen
Ist nach Ihnen wohl die Wohnungsspekulation mit den obigen Antworten zu vereinbaren?

concerned in order to promote the industrialization of building?
Urbanization is a matter of politics: the housing problem and its ramifications stem from it.
Can we some day hope for a public service for which the right to housing has become a duty? In that case, industrialization would have found its market.

8. Which form of industrialization do you consider the most promising?
All industrialized products on a large scale are very manifold; we can conclude from this fact that neither the system nor the elements nor techniques have priority.

9. Remarks
Do you think that housing speculation is compatible with the above replies?

Maurice Silvy

Frankreich, Architekt; tätig auf dem Gebiet des industrialisierten Bauens. Entwicklung von Bausystemen für die Industrie. Forschung auf dem Gebiet des evolutiven industrialisierten Wohnungsbaus.

1. Definieren Sie «die Industrialisierung des Bauens»
Die Industrialisierung des Bauens ist ein Prozeß vollkommener Erneuerung der Wirtschaftlichkeit des Wohnraums im Lichte neuer Bedürfnisse der Gesellschaft sowohl wie des Einzelnen, dank den Konzepten, Methoden und Techniken unseres industriellen Zeitalters.

2. Welches sind die erfolgreichsten Unternehmen des industrialisierten Bauens?
Ich glaube, daß die Modelle von Jean Prouvé qualitativ mit den besten Flugzeugen und Automobilen unserer Zeit verglichen werden können; quantitativ sind Unternehmen wie Levittown und National Home in den Vereinigten Staaten oder die in Rußland entwickelte Massenproduktion ein Erfolg. Man muß aber mit Bedauern feststellen, daß eine praktisch vollkommene Scheidung zwischen Quantität und Qualität besteht.

3. Nach welchen Maßstäben definieren Sie den Grad des Erfolges?
Es gibt viele und verschiedene Grade des Erfolges. Alles hängt vom Standpunkt des Beschauers ab. Handelt es sich um den finanziellen Erfolg eines Unternehmens, das seine Bautätigkeit auf bestimmte Zwecke spezialisiert? — oder um den Erfolg, der den zahlreichen Benutzern naheliegt, so daß sie frei ihre Wahl bestimmen können? Es gibt eine große Anzahl von Maßstäben, und viele sind rein subjektiv. Für mich heißt Erfolg die gezielte Reaktion auf ein sozio-ökonomisches Bedürfnis an einem gegebenen geschicht-

France, architect, active in the field of industrialized building. Has participated in the development of systems for building enterprises. Research on evolutionary industrialized housing.

1. Define "the Industrialization of Building"
The industrialization of building is a process of perfect renewal of the viability of housing in the light of new needs of society and of the individual, thanks to the concepts, methods and techniques of our industrial age.

2. Which have been the most successful industrialized building ventures?
I believe that the models of Jean Prouvé can be compared, qualitatively, with the best airplanes and motorcars of our age; on the quantitative plane, ventures like Levittown and National Home in the United States or the mass production developed in Russia are successes. However, we must face the regrettable fact that there is a complete gap between quantity and quality.

3. By which parameters do you define the degree of success?
There are many, different degrees of success. Everything depends on the standpoint of the observer. Have we to do with the financial success of a concern which specialized its building activity? — or success that bears on the numerous users of the building, so that they can freely exercise their choice? There are a large number of standards, and many are purely subjective. For me success means the purposeful reaction to a socio-economic need at a given point of historical time. At the present time, this reaction must be structurally formative, evolutionary and adapted to the needs of the greatest number of people.

lichen Zeitpunkt. In der heutigen Zeit muß diese Reaktion strukturgebend, evolutiv und den Bedürfnissen der größten Anzahl von Menschen angepaßt sein.

4. Wie unterscheiden sich industrialisierte Bauten von konventionellen Bauten?
Es gibt noch keine echten Verwirklichungen der Industrialisierung. Sogenannte industrialisierte Gebäude werden in gleicher Weise konzipiert, produziert und auf den Markt gebracht wie die konventionellen. Neue Vorschläge hängen von der Evolution der Gesellschaft ab, von ihren entstehenden Bedürfnissen und von den Mitteln, die sie anzuwenden beschließt. Die Veränderung der Formen spiegelt die Veränderung des Geistes.

5. Welche Vorteile bietet die Bauindustrialisierung im Vergleich zu heutigen konventionellen Methoden?
Ein wirklich industrialisierter Bau könnte viele Dienste erfüllen und Ausdruck einer anderen sozialen Ordnung sein als die Bauten von heute. Wenn man die Ziele klar definiert, könnten die technischen Mittel, die uns zur Verfügung stehen, schon jetzt sehr annehmbare Lösungen bieten. Man muß aber feststellen, daß diese Lösungen nicht in isolierten Bauten zu suchen sind, sondern in eigentlichen «habitats». Ein solches echt industrialisiertes «habitat» kann erst geplant werden, wenn die Bedürfnisse im Lichte einer globalen und integrierten Formel definiert sind und nicht in bruchstückhafter Manier, so wie es heute üblich ist.

6. Wie kann der Architektenberuf am besten einen Beitrag zum Industrialisierungsprozeß leisten?
Der heutige Architektenberuf in Frankreich sowie in zahlreichen anderen Ländern ist der Zeit nicht angepaßt. Zu oft bleibt er in Isolation außerhalb der aktuellen Probleme. Die Aufgabe des Architekten liegt in der Schaffung und Förderung einer den Menschen angepaßten Raumordnung. Dieses Konzept muß auf allen Ebenen des Industrialisierungsprozesses formuliert und verteidigt werden. Daher ergibt sich für den Beruf die Notwendigkeit einer explosiven Ausstrahlung in vielen Richtungen (Administration, Unternehmen, Promotion usw.), um sehr weitverzweigte Aufgaben zu erfüllen. Man könnte sagen, daß der Architekt in der Bauindustrialisierung den Humanisten vertreten sollte, während der Ingenieur Vertreter der Technik ist – ohne die Möglichkeit des Architekten-Ingenieurs auszuschließen.

7. Welche Rolle spielt die politische Macht – oder welche Rolle kann sie spielen –, um die Bauindustrialisierung zu fördern?
Die politische Macht muß Dienerin der Gesellschaft sein und nicht ein Selbstzweck, der seine Berechtigung in einigen Prestige-Unternehmen sucht. Es bedarf einer ausgedehnten, vereinten Aktion der Forschung und Experimentierung, um der Industrialisierung zum Aufschwung zu verhelfen. Die politische Macht müßte den Knoten durchschneiden, vor allem, um eine Umgestaltung der Berufsstruktur des Bauens herbeizuführen und die Öffentlichkeit auf breiter Basis zu informieren. Jedoch glaube ich, daß eine echte Industrialisierung des «habitats» nur mit Hilfe einer politischen Macht entwickelt werden kann, die der Mitwirkung des Einzelnen und der verschiedenen Be-

4. In what ways do industrialized buildings differ from conventional buildings?
There are as yet no genuine realizations of industrialized building. So-called industrialized buildings are conceived, produced and put on the market just like conventional ones. New proposals depend on the evolution of society, on its emergent needs and on the means it decides to apply. Alteration of forms mirrors the alterations going on in the mind.

5. What can industrialized building offer that conventional methods cannot provide today?
A truly industrialized building could perform many services and be the expression of a different social structure from that reflected by the buildings of today. If the goals are clearly defined, the technical resources available to us could even now offer very acceptable solutions. However, we must realize that these solutions are not to be sought in isolated buildings, but in residential centres. Such a genuine industrialized residential centre can be planned only if needs are defined in the light of a global and integrated formula, and not in the piecemeal way employed at present.

6. How can the architectural profession best contribute to the industrialization process?
The architectural profession in France at the present time, as well as in many other countries, is not up to date. Far too often it remains isolated from contemporary problems. The task of the architect consists in the creation and promotion of a spatial order adapted to the human being. This concept has to be formulated and defended on all levels of the industrialization process. Therefore the profession needs to expand vigorously in all directions (administration, business enterprise, promotion, research, etc.) in order to carry out widely ramifying jobs. It could be said that the architect ought to be the humanist in the field of industrialized building, while the engineer is the technical man – but we must not exclude the possibility of the architect-engineer.

7. What is or can be the rôle of the political power concerned in order to promote the industrialization of building?
The political power must be the servant of society and not an end in itself seeking its justification in a few prestige ventures. There is needed an extensive, unified campaign of research and experiment to get industrialized building going. The political power would have to cut the Gordian knots, especially in order to bring about a reorganization of the professional structure in the building field and to orient the public on a broad basis. However, I believe that genuine industrialization of housing can only be developed with the aid of a political entity that is far more open to co-participation by private citizens and groupings than is now the case, as we all know, in both West and East.

8. Which form of industrialization do you consider the most promising?
In my opinion, industrialization can really develop only on the plane of a global housing programme. All the formulas we now know are only piecemeal; but they do point in the direction of a new concept: that of the general, large-scale polyvalent building. These evolv-

völkerungsgruppen viel weiter offensteht, als es heute im Westen und Osten bekanntlich der Fall ist.

8. Welche Form der Bauindustrialisierung scheint Ihnen am meisten zu versprechen?
Meines Erachtens kann sich die Industrialisierung nur auf der Ebene eines globalen menschlichen «habitats» voll entfalten. Alle Formeln, die wir kennen, sind nur Stückwerk; doch zeigen sie den Weg zu einem neuen Konzept: das der allgemeinen, weiträumigen Mehrzweckbauten. Diese evolutiven Gebäude können mehrere Ebenen der Integration für verschiedene und mannigfaltige Fertigteile enthalten. Es entstehen neue Konzeptionen: Megastrukturen, «structures d'accueil évolutives», Makro- und Mikrostrukturen ... dieses Konzept öffnet der Bauindustrialisierung weite Perspektiven für das nächste Jahrzehnt.

9. Bemerkungen
Zum Abschluß möchte ich sagen, daß wir keine Wahl haben. Die Industrialisierung des Bauens ist unabwendbar. Sie ist die einzig mögliche Antwort auf die riesigen Baubedürfnisse in allen Ländern der Welt. Es genügt aber nicht, eine rein quantitative Lösung zu finden. Qualitätsforschung ist die wesentliche Vorbedingung für das Überleben unserer Zivilisation. Aber was für eine Qualität? Darüber herrscht noch wenig Klarheit. Der Impuls muß von den Benutzern kommen anstatt von Technokraten, Organisatoren, Industriellen und Architekten.

ing buildings can comprise several levels of integration for different and manifold prefab components. New terms are being coined: megastructures, "evolving recipient structures", macro- and microstructures ... this concept opens up broad horizons for industrialized building in the coming decade.

9. Remarks
In conclusion, I should like to say that we have no choice. The industrialization of building is inevitable. It is the sole possible answer to the enormous building needs in all countries of the world. It is not enough, however, to find a purely quantitative solution. Quality research is the essential prerequisite for the survival of our civilization. But what kind of quality? On this point we are not at all clear. The instigation must come from the users of buildings instead of from technocrats, organizers, industrialists and architects.

Riccardo Meregaglia

Italien, Dipl.-Ingenieur, Mailand. Inhaber der «Impresa Generale Costruzioni MBM, Meregaglia S.p.A.». Partner von «Building System International», tätig in den USA. Vizepräsident von AIP (Italienische Gesellschaft für Vorfertigung). Präsident des Komitees für Vorfertigung von ANCE (Italienische Unternehmer-Gesellschaft).

1. Definieren Sie «die Industrialisierung des Bauens»
«Industrie» bedeutet die Organisation und Planung des Produktionsprozesses — eher maschinell als von Hand — mit dem Ziel, den Markt laufend zu versorgen. Voraussetzung dafür ist eine erfolgreiche Technologie und ein fließender Verkauf. Die Organisation wird selbstverständlich mit der raschen Entwicklung neuer Baustoffe und der neuen Technologie Schritt halten. Die heutigen Technologien bleiben nicht stehen; morgen werden sie neuen Entwicklungen weichen. Und so ist es unumgänglich, ein «alleiniges Management» zu haben, also die Organisation in einem Leitungsstab zu zentralisieren, der alle größeren Probleme, angefangen von der Materialbeschaffung bis zum Verkauf der Fertigprodukte, lösen muß. Folglich muß das Produkt ein «Modell» in seiner Gesamtheit sein. Das Produkt ist also das Haus als Ganzes und nicht seine Elemente.

2. Welches sind die erfolgreichsten Unternehmen des industrialisierten Bauens?
—

Italy, civil engineer, Milan. Owner of the "Impresa Generale Costruzioni MBM, Meregaglia S.p.A.". Partner of "Building System International" operating in the U.S.A. Vice-President of AIP (Italian Prefabrication Association). President of the prefabrication committee of ANCE (the Italian Constructors Association).

1. Define "the Industrialization of Building"
"Industry" means the organization and planning of the productive process aiming at meeting a market of repetitive series through the flow of continuous production implemented by machine rather than by hand. A premise to all this is a good technology and a sales continuity. It is obvious that the organization will follow the quick development of new materials and technology. Today's technologies are temporary; tomorrow they will be different.
To do this it is necessary to have a "single management", that is to centralize the organization in a managerial staff solving all the major problems from the procurement of materials to the sale of finished products. It follows that the product must be a "model" in its complete unity. Therefore the product must be the house as a whole and not its components.

2. Which have been the most successful industrialized building ventures?
—

3. By which parameters do you define the degree of success?

3. Nach welchen Maßstäben definieren Sie den Grad des Erfolges?

Auch wenn die «Produzenten» von Häusern die Dinge etwas anders sehen, so besteht heute kein Zweifel mehr, daß der korrekte Standpunkt beim Konsumenten zu suchen ist. Für den Konsumenten ist es unwichtig, wie oder mit welchen Materialien ein Haus hergestellt wird, es muss nur funktionell, schön, solid und preiswert sein.

Um die Fragen 2 und 3 zu beantworten, wäre es notwendig, eine objektive Klassifikation der vielen Unternehmen aufzustellen, die auf diesen vier Anforderungen gegründet ist.

Die Untersuchung sollte sich meiner Ansicht nach auf jene Fälle beschränken, in denen der Konsument tatsächlich zufriedengestellt wurde, und sollte die beste Art, diesen zu befriedigen, herauskristallisieren. Man darf nicht vergessen, daß der Konsument wenigstens bis jetzt nicht am Bauprozeß beteiligt war.

Ein öffentlicher oder privater Agent wird eingesetzt, um die Wünsche und Geldmittel des Benutzers zu vertreten, der das Produkt annimmt – ob es gut oder schlecht ist –, da ihm keine andere Wahl bleibt.

Die beste Art, einen Konsumenten zufriedenzustellen, scheint heute ein privates Unternehmen zu sein, gemeinsam von einem Promoter und einem industrialisierten Produzenten, nach dem Konsumenten ausgerichtet, der, wenn nötig, von der öffentlichen Behörde durch Beiträge unterstützt wird, um die Hypothekarzinsen zu verringern.

4. Wie unterscheiden sich industrialisierte Bauten von konventionellen Bauten?

Wenn in einem Land zu einer bestimmten Zeit eine große Nachfrage nach Wohnungen und ein Mangel an spezialisierten Arbeitskräften zugleich eintritt, dann kann nur ein Zurückgreifen auf eine industrialisierte Organisation, wie sie unter Punkt 1 definiert wurde, die Situation retten. Wir kennen dieses Phänomen seit den letzten hundert oder hundertfünfzig Jahren in vielen Ländern bei jedem Produkt, das auf dem Markt verlangt wird.

Um die besten Resultate zu erzielen, muß sich das Bauen mit Systemen wie eine tatsächliche Industrie abwickeln. Sie kann nicht den konventionellen Markt annehmen, wie etwa das Feilschen um Planung und Zeitplan eines anderen: sie muß ihr eigenes industrielles Produkt und einen eigenen ständigen Markt entwickeln. Um erfolgreich zu sein, muß dieses Produkt den Ansprüchen und Geldmitteln des Kunden entsprechen, auch wenn die Verteilung durch «Agenten», wie zum Beispiel Wohnbaubehörden, Promoter usw., vor sich geht. Die Forschung ist wichtig, um nicht nur die quantitativen, sondern vielmehr die qualitativen Anforderungen des Konsumenten aufzudecken, und zwar bezüglich der individuellen Behausung wie bezüglich des Städtebaus.

Wir müssen die Behausung wieder neu erfinden.

5. Welche Vorteile bietet die Bauindustrialisierung im Vergleich zu heutigen konventionellen Methoden?

Zuerst bildet sie die einzig mögliche Lösung der oben beschriebenen Situation. Tatsächlich verlangt die Kontinuität, welche die Basis der industriellen Tätigkeit darstellt – im Gegensatz zur konventionellen Bauweise, die sporadisch ist –, eine Marktorientierung. Um

Even if the "producers" of houses see things in a different way there is today no doubt that the correct point of view must be that of the consumer. For the consumer it is of no importance how or with which materials his house is produced, but it must be functional, beautiful, solid and economical.

To answer question 2 and 3, it would be necessary to make an objective classification of the numerous ventures based on these four requirements.

I think the investigation should be limited to the cases where the consumer was really satisfied and to individualize the best way to satisfy him. One must remember that, at least up to now, the consumer has been a non-participant in the building process.

A public or private agent acts to represent the desires and economical means of the user who accepts the product – whether it be good or bad – since he has no other choice.

Today the best way to satisfy the consumer seems to be a private activity carried out jointly by a developer and an industrialized producer directed towards the consumer who, if needed, is helped by the public authorities with contributions for reducing the mortgage interest rates.

4. In what ways do industrialized buildings differ from conventional buildings?

When in a certain country and at a certain time there coexists a big demand for housing and a shortage of specialized manpower, the only way of resolving the situation is to fall back on industrial organization as defined in point 1. This phenomenon has taken place during the last hundred or hundred and fifty years in many countries for every product required by the market.

To give the best results, system building has to act as an actual industry. It cannot accept the conventional market such as bidding on somebody else's design and scheduling: it must develop its own industrial product and market it with continuity. To be successful this product has to meet the consumer's needs and economic capacity, even if the distribution is done through "dealers" such as housing authorities, developers, etc. Research must be carried out in order to know not only the quantitative but especially the qualitative consumer requirements of both the individual dwelling and the urban development.

We have to reinvent the dwelling.

In fact, while using conventional materials and technologies the results are well known and satisfactory; this is no longer true when using new ones.

5. What can industrialized building offer that conventional methods cannot provide today?

First of all, it gives the only possible solution to the above mentioned situation. In any case the continuity which is the basis of the industrial operation requires – to the contrary of the conventional building which is sporadic – market orientation. In other words it must, to achieve success, base its model on market research, analyse the consumer feedback, renounce the plus value of land in order to concentrate on industrial benefits.

6. How can the architectural profession best contribute to the industrialization process?

also des Erfolges sicher zu sein, muß sie mit anderen Worten ihre Modelle auf die Marktforschung abstellen, den Konsumenten-«Feedback» analysieren und den Mehrwert des Bodens aufgeben, um sich auf industrielle Gewinne zu konzentrieren.

6. Wie kann der Architektenberuf am besten einen Beitrag zum Industrialisierungsprozess leisten?

Die konventionelle Bauweise hat die archaische Struktur ihrer Anfänge beibehalten: ein leitender Kopf, der Architekt-Bauer (die Etymologie des Wortes ist «arké» = Meister, «tekton» = Bauer). So ist der traditionelle «ménage à trois» zustande gekommen: Bauherr–Hersteller–Architekt, wobei der letztere eine demiurgische Stellung einnimmt.

Durch die technische, technologische und neuerdings auch organisatorische Evolution wird diese Struktur veraltet, wenn man zur Industrialisierung übergeht.

Die neue Struktur kopiert diejenige der Industrie; der Hersteller gibt sein Produkt direkt auf den Markt, der Architekt arbeitet als Designer mit dem Hersteller zusammen, um den vier Ansprüchen zu genügen, die ein Haus erfordert.

7. Welche Rolle spielt die politische Macht – oder welche Rolle kann sie spielen –, um die Bauindustrialisierung zu fördern?

Die politischen Behörden sollten nach meiner Ansicht – wie es bereits in einigen Ländern erfolgreich geschieht – eine doppelte Rolle spielen. Erstens sollte die Bautätigkeit von den Behörden ohne Einschränkungen gefördert werden. Das bedeutet also, das ganze Gebiet zu rationalisieren, eine Nachfrage zu schaffen, die Nutzung des Bodens zu organisieren, Land zu vernünftigen Preisen zur Verfügung zu stellen, den unteren Schichten mit Hypotheken auszuhelfen und die Kontinuität der Finanzierung zu garantieren. Als Zweites sollte sie die Industrialisierung des Bauens fördern. Viele Länder bewegen sich schon in dieser Richtung. Der beste Weg besteht meiner Meinung nach darin, den notwendigen Markt zu garantieren, um die Errichtung einer industriellen Organisation zu rechtfertigen – die «goldene Regel» ist ein Vertrag von 5000 Einheiten in fünf Jahren –, wobei es dem Hersteller überlassen bleibt, den privaten Markt zu entwickeln.

8. Welche Form der Bauindustrialisierung scheint Ihnen am meisten zu versprechen?

Bis jetzt ging dieses Thema nur die Techniker an, die zum vornherein darauf aus sind, die bestmöglichen technischen und technologischen Lösungen herauszufinden. Doch für jede Industrie liegt das Grundproblem nicht in der Produktion, sondern im Verkauf, was heißt, daß der kommerzielle Aspekt zuerst, also vor dem technischen kommt. Meines Erachtens sollte die Frage nach folgenden Gesichtspunkten gelöst werden:

a) Organisation des Marktes für industrialisiertes Bauen (Kontinuität als wesentlicher Faktor)
b) Organisation der Produktionsstruktur (alleiniges Management wesentlich)
c) Wahl der Technologie, der Form der Industrialisierung, der Grundbaustoffe, des Typus, der Größe und des Ortes des Produktionswerkes, der Werkzeuge usw.

Conventional building has maintained the archaic structure of its origins: one directing mind, the architect-builder (the etymology of the word architect is "arké" = chief, "tekton" = builder).

This has created the traditional "ménage à trois", client – producer – architect with the latter in a demiurgic position.

The technical, the technological, and now also the organizational evolution make this structure obsolete, when one moves onto industrialization.

The new structure copies that of the industries; the producer gives his product directly to the market, the architect collaborates as a designer with the producer to satisfy the four requirements which characterize the house.

7. What is or can be the rôle of the political power concerned in order to promote the industrialization of building?

Ih my opinion the political authorities should – as they effectively do in some countries – play a double rôle.

The first one should be to sponsor the building activity without any kind of qualification. That means to rationalize the whole field, organize the demand, organize the use of the land, make land available at reasonable cost, help the poorer classes through mortgages, and assure continuity of the financing.

The second action, if the condition exists, should be to sponsor industrialized building. Several approaches have been adopted in this direction by some countries. In my opinion the most effective way is to ensure the necessary market to justify the setting-up of the industrial organization – the "golden rule" is a contract of 5000 units in five years – leaving then the producer to develop the private market.

8. Which form of industrialization do you consider the most promising?

The subject has been, until now, monopolized by technicians who are naturally engaged in finding out the best possible technical and technological solutions. But for any industry, the first problem is not production but selling. that is, the commercial aspect comes first, followed by the technological one.

In my point of view the answer to the question should be developed in the following order:

a) Organization of the market for industrialized building (continuity being essential)
b) Organization of the production structure (single management being essential)
c) Choice of technology, of the form of industrialization, of the fundamental materials to be utilized, of the type, size and location of the production factories, of the tools, etc.

Robertson Ward, Jr.

U.S.A., FAIA, architect, research and system consultant, engaged since 1950 in industrialization, has worked with Konrad Wachsmann, Luigi Nervi, was head of Research and Development for Skidmore, Owings & Merrill, independent systems work includes design and development of major SCSD California Systems, chairman of AIA National Housing Technology Committee, chairman of AIA Operation Breakthrough Review Committee.

1. Define "the Industrialization of Building"
The rational ordering of the process of matching our contemporary physical resources (materials, production, transport and assembly techniques, labour's manpower and skills, the realities of material, economic and governmental structures) with the identified social needs to achieve the optimally balanced human habitat and activity shelter. This balanced resource allocation must utilize a scope and scale which encompasses the more sophisticated and universal organizations and techniques of industry.

2. Which have been the most successful industrialized building ventures?
Most "successful" equals "with lessons to be learned".
a) The Crystal Palace
b) The pre-engineered Industrial Metal Building
c) Classic lessons: General Panel Corp., CLASP, Allside Homes, Larson & Nielsen, Fritz Haller, Marburg, the Mobile Home.

3. By which parameter do you define the degree of success?
The clarification of the integrating disciplines involved: i.e., geometric, modular and topological choices, the choice of limited, but adequate, permutations for specific social uses, the solution of problems by the rigorous awareness of how to eliminate unnecessary problems, the opportunities of technical scale and organizational scope, the more comprehensive allocation of planning and design efforts, the anticipation of problems of choice, flexibility and change.

4. In what ways do industrialized buildings differ from conventional buildings?
Organization of social and technical energies towards a continuity of need and solution, thereby diminishing duplication and waste of materials, of labour, of time, of organizational, planning and social resources.

5. What can industrialized building offer that conventional methods cannot provide today?
Industrialized building can (only in its best examples) offer: predictability, accuracy, controlled performance, choice within limited permutations, flexibility for change, higher performance for equal investment, economy in time and effort.

6. How can the architectural profession best contribute to the industrialization process?
In key rôle as multi-discipline synthesizers; design is the action-verb in the matching of technical and

In Schlüsselpositionen als Förderer von Synthesen vieler Disziplinen; Gestaltung stimmt die technischen und materiellen Schätze auf soziale Bedürfnisse ab. Der wichtigste Maßstab ist die sozial relevante Ausnützung der technischen Möglichkeiten.

7. Welche Rolle spielt die politische Macht – oder welche Rolle kann sie spielen –, um die Bauindustrialisierung zu fördern?
Am wichtigsten wäre es, die Kommunikations- und Berechnungsgrundlage für soziale Kriterien, Bedürfnisse und Auswirkungen auf das gesamte menschliche Habitat zu schaffen, um die spezifischen, physikalischen und technischen Lösungen beurteilen zu können: Organisation für ein Optimum an sozialer Auswahl, und vor allem Aktion!

8. Welche Form der Bauindustrialisierung scheint Ihnen am meisten zu versprechen?
Die Entwicklung von wirkungsvollen investitionsintensiven Werkzeugen für die Baustellentechnik könnte sich als ebenso wichtig erweisen wie die Vorfertigung oder die Entwicklung von vorgefertigten Komponenten. Ohne Zweifel werden sich verschiedene Komplikationen auf und außerhalb der Baustelle, in der Montage und in der lokalen, regionalen, kontinental und interkontinental zentralisierten Produktion einstellen. Organisationsauswertung und Kommunikationstechniken werden äußerst wichtig sein.

9. Bemerkungen
Können wir von «reinen» System-Leistungsbeschreibungen sprechen? Wir sollten bei allen Arten von Leistungen möglichst spezifisch sein, doch sind es nur die einfachen Charakteristika, die in «reiner» Form beschrieben werden können. Sobald wir Leistungen kombinieren (wie ein Modul und eine Beleuchtungsebene, eine Vielzahl von Flexibilitätsmöglichkeiten und eine feuertechnische oder akustische Trennung usw.), haben wir das Spektrum von vorhersagbaren Lösungen eingeschränkt und einen bewußten Schritt in Richtung einer Designsynthese unternommen. Nach meiner Meinung könnten wir die Systemleistungsentwicklung folgendermaßen beschreiben: Sie entspringt teilweise einem umfassenden Leistungsbedarf, bewegt sich aber auch entlang eines zunehmend konvergierenden und iterativen Leistungsspektrums von Wahlmöglichkeiten und befindet sich auf diese Weise bereits auf dem Weg zu einer endgültigen Systemsynthese. Der Prozeß der Synthese hat bereits begonnen, bevor die Aussage bezüglich der «Leistung» vollendet war.

physical resources to social needs. The key judgment is the social efficacy of technical possibilities.

7. What is or can be the rôle of the political power in order to promote the industrialization of building?
Most necessary to establish the communication and evaluatory base of social criteria, needs, effects on all human habitat scales against which the specific, physical and technical solutions can be judged: organization for optimum social choices, and action!

8. Which form of industrialization do you consider the most promising?
Developing effective high investment-intensive site technique tools may be as important as any prefabrication, component factory-produced systems development. Undoubtedly, various complications of in-situ, on-site process and assembly, and local, regional, continental and intercontinental centralized production will apply. Organization, evaluation, communication techniques will be paramount.

9. Remarks
Can we write "pure" system performance requirements? We should be as specific as we can on all areas of performance, however, it is only the relatively simple characteristics which can be "purely" described. As soon as we start combining performances (i.e., a module and illuminating level, a multiple of flexibility choices and a fire or acoustic separation, etc.) we have narrowed the spectrum of predictable solutions and have begun a conscious step towards a final design synthesis. I believe we should more accurately describe a system's performance development as originating partly in a broad performance, but progressing, iteratively, along an increasingly converging performance spectrum of choices, already headed towards a final systems synthesis. The synthesis process has already begun before the "performance" statement can be completed.

Gyula Sebestyén

Ungarn, Professor, Direktor des Institutes für Bauforschung, Budapest.

1. Definieren Sie «die Industrialisierung des Bauens»
Die wichtigste Bedingung für die Industrialisierung des Bauens liegt außerhalb der Bauindustrie; man braucht moderne Baustoffe, Komponenten und Maschinen.

Hungary, professor, director of the Institute for Building Science, Budapest.

1. Define "the industrialization of building"
The most important condition for the industrialization of building lies outside the construction industry; up-to-date building materials, components and machines are needed.

2. Welches sind die erfolgreichsten Unternehmen des industrialisierten Bauens?
a) Methoden, Wohnbauten mit großen, vorgefertigten Betonpaneelen zu bauen
b) «Mobile Homes» und Zellen
c) Dreidimensionale Einheiten.

3. Nach welchen Maßstäben definieren Sie den Grad des Erfolges?
Massenproduktion, hohe Arbeitsproduktivität, hohe technische Leistung.

4. Wie unterscheiden sich industrialisierte Bauten von konventionellen Bauten?
Durch allgemeine gute Qualität und durch bessere Vorbedingungen, um die Nachfrage befriedigen zu können.

5. Welche Vorteile bietet die Bauindustrialisierung im Vergleich zu heutigen konventionellen Methoden?
Ein quantitativ großer Ausstoß der Bauindustrie.

6. Wie kann der Architektenberuf am besten einen Beitrag zum Industrialisierungsprozeß leisten?
Indem er die Möglichkeiten diverser moderner Industriesektoren maximal ausnützt.

7. Welche Rolle spielt die politische Macht — oder welche Rolle kann sie spielen —, um die Bauindustrialisierung zu fördern?
Durch die Organisation des Informationsdienstes; durch finanzielle Beiträge an technische Experimente und Prototypprojekte.

8. Welche Form der Bauindustrialisierung scheint Ihnen am meisten zu versprechen?
Die Leichtbauweise.

2. Which have been the most successful industrialized building ventures?
a) Methods to build flats with large precast concrete panels
b) Mobile homes and "modules"
c) Three-dimensional units.

3. By which parameters do you define the degree of success?
Mass-production, high productivity of labour, high technical performance.

4. In what ways do industrialized buildings differ from conventional buildings?
By their standard good quality and better conditions to meet the demand.

5. What can industrialized building offer that conventional methods cannot provide today?
A high quantitative output of the construction industry.

6. How can the architectural profession best contribute to the industrialization process?
In making maximum use of the possibilities of various up-to-date industrial sectors.

7. What is or can be the rôle of the political power concerned in order to promote the industrialization of building?
By organizing the information service; by financial contributions to technical experiments and pilot projects.

8. Which form of industrialization do you consider the most promising?
The use of lightweight structures.

7. Zukünftige Entwicklungen

7.1 Technologische und technische Entwicklungen

Gegenwärtig hat die Bauindustrialisierung ein befriedigendes Niveau in der technologischen Entwicklung, was die Werkproduktion von Komponenten mit klassischen Baustoffen betrifft.

Es ist schwierig, größere Entwicklungen in der Produktion von Stahl- und Betonstrukturen oder Sandwichpaneelen (ob aus Beton, Stahl, Aluminium oder Sperrholz usw.) vorauszusehen. Auch die Fugenverbindung und die Dichtungstechnik sind zufriedenstellend gelöst, und einige Standardlösungen werden – mit geringfügigen Abweichungen – praktisch in allen wichtigeren industrialisierten Bausystemen dargeboten.

Die beiden offensichtlichsten Schwächen in der Bauindustrialisierung (vorgefertigte wie modulare Systeme) bilden gegenwärtig die Trennwandsubsysteme und die mechanische Ausrüstung. Heute werden Dutzende von Subsystemen auf dem Markt angeboten, und zwar feste wie demontable; alle aber versagen in zwei Punkten: in den Kosten und im Schallisolationswert.

Die von den Herstellern verfügbaren Daten über den Schallisolationswert sind kaum verläßlich. Teste im Labor werden unter streng kontrollierten Montagebedingungen ausgeführt, und es werden oft hochbefriedigende Schallreduktionen gemessen (40–50–55 db).

Auf dem Bauplatz sind die Kontrollen verständlicherweise weniger streng, und die Schallumgehung oder undichten Stellen werden so häufig, daß sich die Reduktionsverluste gewöhnlich um 20 db bewegen. Eine Untersuchung, die EFL in etwa fünfzig Schulen in den USA unternommen hatte, erbrachte den Beweis, daß die durchschnittliche Schallreduktion um annähernd 15–20 db niedriger war als die von den Herstellern angegebene. Die Kosten des vorgefertigten Trennwand-Subsystems (besonders die demontable Variante, aber auch die feste) sind in den meisten Ländern größer als die Kosten der Trennwände, die mit der konventionellen Bauweise errichtet werden.

Eine interessante Lösung bildet vielleicht das örtliche Spritzen der Trennwände (mit dem Vorteil, jegliche Art von Umgehungen und undichten Stellen zu beheben) und die einfache Umgehung des Problems durch vollkommene schalldämpfende Decken und Böden bei den demontablen Trennwänden. Wie immer die Lösung aussehen mag, die Bauindustrialisierung

7. Future Developments

7.1 Technological and technical developments

The present situation of the industrialization of building shows a satisfactory level of technological development in connection with the factory production of components with classic materials.

It is difficult to foresee major developments in the production of steel and concrete structures or sandwich panels (be they out of concrete, steel, aluminium or plywood, etc.). The problem of jointing and sealing can also be taken as being satisfactorily solved, and a few standard solutions, with minor variations, can practically be found in all major industrialized building systems.

The two most serious weaknesses of present industrialized building (prefabricated or modular system) are the partition sub-systems and the mechanical equipment. Today we can find dozens of partition sub-systems on the market, fixed or demountable; but all of them fail on two points: cost and sound-insulation value.

On sound-insulation value the data made available by the manufacturers are highly unreliable. Laboratory tests are conducted under severely-controlled assembly-conditions and often highly satisfactory sound-reduction rates are achieved (45–50–55 db).

On the building site the controls are logically less stringent and the outflanking or leakages become so important that losses in the reduction rate of some 20 db are common. A study conducted by EFL in some fifty schools in the United States proved that the average sound-reduction value was by approximately 15–20 db inferior to that claimed by the manufacturer. The cost of the prefabricated partition sub-system (particularly for the demountable kind, but also for the fixed one) remains in most countries greater than the cost of partitions by conventional methods. An interesting solution may well be the site-spraying of the partition (with the advantage of avoiding any kind of leakage and most outflanking) for the fixed kind, and in the simple by-passing of the problem by highly absorbent ceilings and floors, for the demountable partitions. Whatever the solution may be, we have here a serious problem for the industrialized building.

The whole field of mechanical services (heating, ventilation, cooling, water provision and disposal, waste disposal, lighting, etc.) is today in a pitiful stage. In housing the utilization of some panel-integrated circuits has indicated a possible solution, but for more

179

steht hier einem sehr schwierigen Problem gegenüber. Der ganze Bereich der mechanischen Einrichtungen (Heizung, Ventilation, Kühlung, Wasserversorgung und Wasserableitung, Abfallbeseitigung, Beleuchtung usw.) ist heute in einem erbärmlichen Zustand. Im Wohnungsbau hat die Neuerung, Leitungen direkt in die Paneele zu integrieren, eine mögliche Lösung aufgezeigt; für komplexere Bauwerke, wie Bürobauten, Schulhäuser und Spitäler, ist jedoch die heutige Situation vollkommen unannehmbar.

Eine gewisse Straffung im Wirrwarr der mechanischen Subsysteme muß erreicht werden. Zieht man das Verhältnis zu den Gesamtkosten des Baus (zwischen 20 und 40%) in Betracht, dann könnte die Industrialisierung dieses Subsystemes ohne weiteres einen größeren wirtschaftlichen Durchbruch für das Konzept der Bauindustrialisierung bedeuten.

Die oben erwähnten Entwicklungen können als unmittelbare Ziele betrachtet werden, ohne wesentliche Einwirkung auf die gegenwärtige Form und Funktion eines Gebäudes. Tiefgreifendere Folgen haben vielleicht Entwicklungen in der Demontierbarkeit und Transportleichtigkeit der Bauwerke selbst und sogar ganzer Stadtteile. Die offensichtliche Notwendigkeit einer solchen «zeitlichen Flexibilität» ist bis jetzt noch nicht eindeutig erwiesen, doch gerade die Beweglichkeit unserer Gesellschaft deutet unweigerlich auf ein wachsendes Interesse für diese Eigenschaft hin. Das Kriterium der «zeitlichen Flexibilität» wird das heutige Grundmerkmal eines Gebäudes (eine statische Struktur) modifizieren und es zu einer dynamischen Struktur umwandeln.

Diese Anforderungen existieren bereits für Schulen, öffentliche Gebäude, Bürobauten und Spitäler; um aber zur vollen Wirkung zu gelangen, müssen diese Anforderungen auf das gesamte «Stadtsystem» zutreffen.

In ähnlicher Richtung geht auch die Entwicklung, die es dem Bewohner eines Wohnkomplexes erlaubt, in den Lageplan und in die funktionale Raumaufteilung aktiv einzugreifen. Diese Anforderung verlangt eine «mobile» maschinelle Ausrüstung, Pakete von raumbildenden Einheiten (die den Bewohnern zur Verfügung gestellt werden), und beeinflußt vielleicht auch den Entwurf des Baus und das Äußere (wie bei der Takara Beautillon, von Kurokawa für die Weltausstellung in Osaka entworfen).

Wir müssen selbstverständlich auch die potentielle Entwicklung der «Kunststoffe» erwähnen. Bis heute haben zwar nur wenige Kunststoffe ihren Weg in die Bauindustrie gefunden. Diese Situation bleibt sich voraussichtlich gleich, doch die Möglichkeit eines Durchbruchs kann nicht ausgeschlossen werden.

Nach meiner Auffassung liegt die Möglichkeit einer durchgreifenden Neuerung eher im Bereich der örtlich produzierten Bauwerke. Der Industrialisierungsprozeß hat bis jetzt versucht, die Kettenproduktionstechnik der Autoindustrie und ähnlicher «Kleinprodukte» zu übernehmen. Es scheint kaum folgerichtig zu sein, daß dieselbe Technik die richtige für eine Industrie ist, die mit großen und kleinen (kleiner Wert im Verhältnis zum Volumen) Produkten handelt. Die Möglichkeit der Intervention komplexer beweglicher Ausrüstung auf dem Bauplatz, die fähig ist, den Bau und/ oder die Schale des Baus zu produzieren, ist wirtschaftlich attraktiv. Bis jetzt sind die Resultate nicht gerade

complex buildings, like offices, schools, and hospitals, the present situation is utterly unacceptable.

A certain streamlining of the mechanical sub-system tangle will have to be achieved. Taking into consideration the incidence on the total cost of the building (from 20 to 40%) the industrialization of this sub-system may well prove a major economical breakthrough for the industrialized-building concept.

Both above-mentioned developments can be considered as short-range objectives, which will not change in depth the present appearance and working function of the building.

More far-reaching may be developments in the demountability and transportability of the building itself, and even of complete areas of the town.

The real necessity for this requirement of "flexibility in time" has, as yet, not been proved beyond all doubt, but the very mobility of our society strongly indicates a growing interest in this feature.

The requirement for "flexibility in time" will modify the present essence of a building (a static structure) and transform it into a dynamic structure.

Requirements of this kind are already in existence for schools, public buildings, offices and hospitals, but in order to become meaningful these requirements will have to be extended to the complete "town system". Following similar lines the requirement which allows the inhabitant of a housing-complex a major intervention in the lay-out and the functional distribution of the space, may become relevant. This kind of requirement will call for "movable" mechanical equipment, packages of space dividers (to be available for use in the course of time by the dwellers), and may even condition the design of the structure and the envelope (as shown in the Takara Beautillon, designed by Kurokawa for the Osaka World Fair).

As usual we cannot forget the potential development of the "plastics". Until today very few plastics have found great success in the building industry. It may be that the situation will not change in the foreseeable future, but the possibility of a break-through cannot be excluded.

I personally see the possibility for a major development in the area of site-produced buildings. Until today the industrialization process has attempted to copy the chain-production approach of the car industry and similar "small-size" commodities. It does not seem logical that the same approach is the correct one for an industry dealing with large and poor (little value for volume) products. The possibility of the intervention of complex, movable equipment on the building site, capable of producing the structure and/or the shell of the building is economically attractive. The results obtained until now are not particularly exciting, but a break-through may well be in the offing, in combination with the development of new materials.

In the factory the introduction — slow, for the moment — of the electronic computer-controlled production-line, may cause a major revolution in the whole concept of industrialized building.

Today we attempt to standardize the components by means of modular coordinated dimensions or by simple prefabrication techniques. The reason is mainly that any change in the dimension of the components requires an adjustment of the jig, mould, cutting tool, etc. with a resulting increase in time and cost.

glorreich ausgefallen, aber ein Durchbruch liegt heute in Verbindung mit der Entwicklung neuer Baustoffe nahe.

Die vorläufig langsame Einführung einer elektronisch computergesteuerten Produktionslinie in die Werkhalle könnte eine tiefschürfende Umwälzung in der ganzen Auffassung der Bauindustrialisierung hervorrufen.

Heute versucht man, die Komponenten durch modulare koordinierte Dimensionen oder durch einfache Vorfertigungstechniken zu standardisieren. Der Grund liegt hauptsächlich darin, daß jede Veränderung der Dimensionen der Bauteile eine Anpassung der Bohr-, Gieß-, Schneidwerkzeuge usw. nach sich zieht, was Zeitaufwand und Kosten vergrößert.

Die elektronisch gesteuerte Maschine vollzieht solche Änderungen ohne bemerkenswerten Zeitunterschied in der Produktionsquote. Daraus resultieren «optimalisierte Komponenten» für eine spezifische Situation und eine Verminderung des gegenwärtigen Materialverschleißes. Der Materialverschleiß entsteht bei einer statischen Dimensionierung, die wegen Sicherheitsfaktoren immer größer als notwendig ist, wie etwa bei Säulenprofilen und Paneel-Befestigungsvorrichtungen, die aus Rücksicht auf Standardisierung für die härtesten Situationen entworfen werden. Ein computergesteuerter Ablauf der Arbeitsvorgänge, der die oben erwähnte Flexibilität berücksichtigt, wurde Mitte der sechziger Jahre im Allside-House-Experiment eingeführt und ist gegenwärtig für die Herstellung von Fenstern erhältlich.

Eine Rückkehr, sei sie zum Vor- oder Nachteil, zum individuell entworfenen Wohnbau kann dann durch die Ausnützung eines höher entwickelten technologischen Werkzeuges vollzogen werden.

Eine weit hergeholte, aber nicht undenkbare Möglichkeit könnte im Do-it-yourself-Wohnpaket liegen. Diese Lösung, die wiederum die Entwicklung von neuen, handlichen Baustoffen nach sich zieht, bekommt vielleicht von daher starke Impulse, daß den meisten Arbeitstätigen immer mehr Zeit zur Verfügung steht, und könnte mit einem «Wegwerf»-Konzept der Wohnung selbst verbunden werden.

7.2 Organisatorische Entwicklungen

Es ist schwierig, wichtige organisatorische Entwicklungen innerhalb der Industrie vorauszusehen. Praktisch haben bereits alle organisatorischen Werkzeuge und Konzepte ihre Anwendung in industrialisierten Bauunternehmen gefunden (auch wenn sie oft mißbraucht werden). Eine tiefgreifende Änderung kann nur dann erfolgen, wenn die Arbeiter direkt am Entscheidungsprozeß mitwirken können, was aber eine politische und nicht eine organisatorische Angelegenheit ist.

Größere Umwälzungen werden gewiß außerhalb der Industrie vor sich gehen müssen.

Die erste, die sich noch innerhalb der Firma abspielt, bildet die Änderung in der Haltung, die wir in Kapitel 1 und 4 skizziert haben. Der «industrialisierte Unternehmer» wird sich immer aktiver an den Markt wenden und das Produkt als eine bereits fertige Ware vertreiben. Eine zweite und sogar wichtigere Änderung betrifft die Stellung des «Benutzers».

The electronic-controlled tool realizes these changes with no noticeable time-differences in the production rate. The result will be "optimalized components" for a specific situation, reducing the present waste of materials caused by a statistical optimization (which, because of safety-factors, is always by excess, as is the case with columns sections, panel-fixing devices, etc., which are, because of standardization preoccupation, designed for the hardest possible conditions). A computer-controlled production line, allowing the above described flexibility, was introduced in the Allside-House experiment in the middle sixties and is presently available for window manufacturing.

A return (for better or worse) to the individually-designed dwelling may then be achieved through the utilization of a more sophisticated technological tool.

A more far-fetched but not impossible development may well be the do-it-yourself dwelling-package. This solution, which again requires the development of new, easily manageable materials, may receive a considerable impulse because of the increased free time available to most working people, and may be coupled with a "disposable" concept of the dwelling itself.

7.2 Organizational developments

It is difficult to foresee major organizational developments inside the industry. Practically all organizational tools and concepts have already found their way into the industrialized building enterprise (even if they are often misused).

A considerable change may take place only through the direct intervention of the workers in the decision-making process, but this is a matter of political, rather than organizational concern.

Major development, however, will certainly have to take place outside the industry.

The first one, still connected with the company itself, is the change of attitude already outlined in chapter 1 and 4.

The industrialized builder will take a more and more active attitude towards the market and make available the product as an already finished commodity. A second and more relevant change will be concerned with the position of the "users".

Im konventionellen Bauprozeß ist der «Benutzer», auch in mehrstöckigen Wohnsiedlungen, immer noch eine leicht identifizierbare Person, die (zum Ärger des Bauunternehmers) Änderungen und Anpassungen der Wohnung auf Grund ihrer Ansprüche verlangen kann. Die Bauindustrialisierung hingegen tendiert auf Grund des quantitativen Faktors dahin, den Benutzer als eine «statistische» Größe abzustempeln, dem jegliches Eingreifen versagt wird. Eine solche Situation ist aus menschlichen und sozialen Überlegungen nicht annehmbar; es muß eine Form der Intervention seitens des Benutzers entwickelt werden. Einfache Untersuchungen über «Anforderungen des Benutzers» genügen nicht (siehe 7.3). Die Öffentlichkeit muß in die Gestaltung einer Umwelt eingreifen können, die auf Grund ihres Ausmaßes das Leben über Jahrzehnte hinaus formen wird.

Doch auch die Abschiebung der Verantwortung auf die Behörden stellt keine zufriedenstellende Lösung dar. Der zukünftige Bewohner selbst muß sich am Spiel beteiligen können. Ohne solche direkte Kontrolle könnte die Industrialisierung des Bauens leicht in ein gefährliches Werkzeug in den Händen der Spekulanten oder, noch schlimmer, in den Händen der Behörden ausarten, und zwar durch den Versuch, der Bevölkerung eine bestimmte Lebensform aufzuzwängen.

Eine dritte und äußerst praktische Entwicklung muß sich im gegenwärtigen System der Kontrollen, Bewilligungen und Untersuchungen abspielen, welche von den Behörden für den Bau verlangt werden. Diese Maßnahmen hatten wahrscheinlich früher einmal die würdige Funktion, die Qualität und Sicherheit eines Bauwerkes zu kontrollieren, um auf diese Weise die Interessen des Benutzers zu schützen. Leider ist dieses System zu einer Zeit in Kraft getreten, als nur wenige Bauten errichtet wurden und es ein paar Jahre dauerte, bis ein mittelgroßes Einfamilienhaus gebaut war.

Heute ist das bürokratische Kontrollsystem zum größten Hindernis der Bauindustrialisierung (und der konventionellen Bauweise) geworden.

Eine kürzlich gemachte Erfahrung möge diesen Punkt unterstreichen. Vor einigen Monaten sollte ich einen kleinen Kindergarten für eine Gemeinde in einem Außenbezirk von Turin entwerfen. Der Entwurf, der zusammen mit den behördlichen Stellvertretern ausgearbeitet wurde, beanspruchte drei Tage. Die voraussichtliche Bauzeit wurde auf sechs bis acht Wochen geschätzt. Durch politischen Druck kam eine Lokalfirma für die Kosten auf, das Geld stand sofort zur Verfügung. Das Projekt ist in der bürokratischen Maschinerie verschwunden. Das Beispiel ist deshalb besonders aufschlußreich, weil sich solche Fälle immer und überall wiederholen.

Der Entwurf eines Schulhauses in Spanien beansprucht zum Beispiel einen Monat, der Bau mit industrialisierten Methoden etwa vier Monate, aber die Baubewilligung, die zahlreiche bürokratische Stellen passieren muß, läßt mindestens zwei Jahre auf sich warten. Unter solchen Bedingungen ist es selbstverständlich müßig, von den zeitsparenden Vorteilen der Industrialisierung zu sprechen.

Immerhin hätte das lange Warten auf die Baubewilligung vielleicht noch einen Sinn, wenn während dieser Zeit dem Projekt tatsächlich eine Verbesserung widerfahren würde. Leider geschieht in dieser Beziehung

With the conventional building-process the "user", even for multi-storey housing projects, is still an easily-identified person who (to the annoyance of the builder) can ask for changes and adaptations of the dwelling to his own particular requirements.

The industrialization of building, due to the quantity factor, tends to transform the user in a "statistical" entity with no possibility of intervention. This situation is clearly unacceptable for social and human reasons and some form of intervention by the user will have to be developed. Simple "user-requirements" studies (see 7.3) are not enough. The public must be able to intervene in the design of an environment which, because of its size, is going to condition its life form for many years to come.

Also "passing the buck" to the public power is hardly a satisfactory solution. The future inhabitant himself must be involved in the game. Without this kind of direct control the industrialization of building may well become a dangerous tool in the hands of speculators or even worse, in those of the public power (attempting to impose a certain life-form onto the population).

A third and very practical development will have to take place in the existing system of controls, permits, checks, which the public administrators put in the way of the construction of any kind of building. Most likely these bureaucratic measures had in their own time the admirable function of controlling the quality and the safety of the building, safeguarding in this way the interests of the user. Unfortunately the whole system was designed at a time when very few buildings were erected and when it took a couple of years to build a modest-sized one-family dwelling.

Today the bureaucratic control system has become the major stumbling block in the way of industrialized building (and of conventional ones for that matter). A recent experience may underline the point.

Some months ago I was asked to design a small kindergarten for a municipality on the outskirts of Turin. The design, made in collaboration with the representative of the municipality, took three days. Expected building time was six to eight weeks. Financing was obtained through political pressure from a local industry, and immediately available.

The project has disappeared in the bureaucratic machinery. This example would not be relevant if it were not a typical case of what occurs daily everywhere.

The design of a school in Spain may take about a month, building it with industrialized methods about four months and to have the project approved by all the involved bureaucratic authorities will take at least two years. Under such conditions talking of the time-saving advantages of industrialization becomes a bad joke.

Even so the considerable time spent by a project while it is under the scrutiny of the public powers would be meaningful, if some kind of improvement on the project were achieved. Unfortunately nothing of this kind takes place. The time is spent in transfering a bunch of drawings and documents from one overworked bureaucrat's desk to the next one until they finally return into the hands of the first. While the whole set of controls, permits, etc., is simply ludicrous, the entire process of public bidding becomes dangerous and at times clearly dishonest. In most countries the bidding

nichts. Die Zeit wird dazu verbraucht, ein Bündel von Zeichnungen und Dokumenten von einem überbeanspruchten Bürotisch zum nächsten abzuschieben, bis die ganze Sache schließlich wieder auf dem ersten Tisch gelandet ist. Während die vielen Kontrollen, Baubewilligungen usw. einfach lächerlich sind, wickelt sich der ganze Prozeß des Ausschreibens auf gefährliche, manchmal sogar auf unehrliche Art ab. In den meisten Ländern verlangen die Ausschreibebedingungen, daß einige Submissionsangebote vorliegen müssen, um eine echte Konkurrenz zu gewährleisten. Aber dieselbe Behörde braucht auch ein bewilligtes Projekt, das von einem Architekten unterzeichnet ist, um den Wettbewerb zu eröffnen. Nun ist es allgemein bekannt, daß jede industrialisierte Methode spezifische Eigenschaften hat, die den architektonischen Entwurf beeinflussen.

Es kommt dann zur folgenden Prozedur:
a) Ein Architekt entwirft das Bauwerk mit der industrialisierten Lösung X.
b) Die Behörde schreibt das Projekt aus.
c) Selbstverständlich kann nur der Unternehmer von Lösung X dem Entwurf und den spezifischen Anforderungen des Projekts genügen.
d) Einige mutige (oder naive) Unternehmer, neben X, machen ein Angebot.
e) Der mutige (oder naive) Unternehmer gewinnt.
f) Das Projekt muß dann abgeändert werden, um den neuen technischen Eigenschaften (in welchem Fall wir wieder bei a angelangt sind) zu genügen, oder ein abgekartetes Spiel zwischen Gewinner und X macht die ganze Prozedur lächerlich.

(Manchmal hat sich X natürlich schon vorher mit seinen Konkurrenten geeinigt, und die Situation ist dann unter Kontrolle.)
Wichtig ist aber, daß die Einführung von industrialisierten Lösungen in den öffentlichen Wettbewerb meistens heißt, eine Gefängnisstrafe zu riskieren. Zum Glück haben einige Behörden die Absurdität einer solchen Lage eingesehen und spezielle Maßnahmen für den Wettbewerb mit industrialisierten Systemen eingeführt. Eine vernünftige Methode stellt zum Beispiel der Vorgang «design and construction» dar. Hier wird den Unternehmern nur eine allgemeine Skizzierung des Bauwerkes unterbreitet, und zwar in Verbindung mit äußerst detaillierten Leistungsbestimmungen. Die Unternehmer können dann den allgemeinen Entwurf den Besonderheiten ihrer spezifischen Bauweise anpassen, und eine Wahl aller Submissionsangebote kann dann vorgenommen werden.
Diese neuen Methoden werden jedoch lediglich bei einigen größeren Projekten angewendet und üben auf die große Mehrheit der Bauwerke keinen wesentlichen Einfluß aus.
Beide Aspekte (Kontrolle und Form der Ausschreibung) müssen sich radikalen Änderungen unterziehen, wenn in den kommenden Jahren dem Bedürfnis nach einer größeren Anzahl Wohnungen entsprochen werden soll. Sonst können wir zusehen, wie alle technischen Verbesserungen unter einem absurden bürokratischen System zerbröckeln oder (was vielleicht noch schlimmer ist) wie die Bürokratie auf der Grenzlinie zwischen legalen und illegalen Praktiken schreitet, um der öffentlichen Nachfrage zu genügen.
Eine wirkliche Verbesserung in unseren operationellen Formen kann vielleicht erreicht werden, wenn die

regulations require that at least a certain number of tenders are received in order to insure proper competition.
But the same authority also needs an approved project, signed by an architect, in order to open the bidding process.
Now it is well known that each industrialized solution has very specific characteristics, which have an influence on the architectural design. We see then the following process taking place:
a) An architect designs the building utilizing the industrialized solution X.
b) The authority asks for tenders from any qualified builder.
c) For obvious reasons only the owner of the solution X can comply with the design and the specifications of the project.
d) Some courageous (or ignorant) builders, beside X, participate in the competition.
e) The courageous (or ignorant) builder wins.
f) The project has then to be changed to comply with its technical capabilities (in which case we return to a) or a private agreement between the winner and X takes place, making a joke of the whole system.
(Clearly, at times, X has reached an agreement with his competitors beforehand, and the situation is then under control.) The point is that to introduce industrialized solutions into the public bidding process often means to risk imprisonment. Fortunately some authorities have seen the absurdity of such a procedure and have introduced special measures for bidding by industrialized systems. A sensible method is for example the so-called "design and construction" approach. Here only a general outline of the building is presented to the contractors in conjunction with a very detailed performance specification.
The contractors can then provide the general design according to the characteristics of their specific building method, and a selection will be made from all designs tendered. However, these new methods are only utilized for a few major projects and have no real impact on the vast majority of buildings.
Both aspects (control and bidding form) will have to undergo radical changes if the need for more dwellings is to be satisfied in the coming years, under the penalty of seeing all technical improvements corroded by an absurd bureaucratic system, or (and maybe worse) of seeing the bureaucrats accept, in order to satisfy the public demand, to operate on the boardline of illegal practices.
A serious improvement in our operational standards may be achieved when governments stop fussing around with amusing and useless technical requirements and come to grips with the problem of land-availability. So long as the present shameless speculation on urban und rural land is allowed to continue any economical or qualitative advantage we obtain by new technical and architectural developments is nullified by the intervention of one single land-speculator.
At times it seems that governments encourage the development of new technologies as a way of preventing the public from focusing onto the real problem. A case in which the above doubt may be raised is that of the U.S.A. with the project called "Operation Breakthrough". Here the United-States government

Regierungen nicht mehr mit veralteten und nutzlosen technischen Anforderungen aufwarten und endlich mit dem Problem des Baulandes fertig werden. Solange man der gegenwärtigen schamlosen Spekulation auf städtischem und ländlichem Gebiet freien Lauf läßt, wird jeder wirtschaftliche und qualitative Vorteil, der durch neue, technische und architektonische Entwicklungen gewonnen wird, durch einen einzigen Bodenspekulanten zunichte gemacht.

Manchmal scheint es, als ob die Regierungen die Entwicklungen neuer Technologien ermutigen, um das Volk vom richtigen Problem abzulenken. Ein solcher Fall ist wahrscheinlich das Projekt «Operation Breakthrough» in den USA. Hier hat die Regierung der USA versucht, die Bauindustrie zu ermutigen, neue Bauweisen aufzugreifen und damit zu experimentieren. Keiner wird sagen wollen, die amerikanische Bauindustrie brauche keine frischen Anregungen, doch es ist eine Illusion, die Probleme der amerikanischen Slums oder der sich ständig ausdehnenden leblosen Vorstädte mit neuen Techniken lösen zu wollen.

Es ist allgemein bekannt, daß nur die hartnäckigen Interessen der Bodenbesitzer und nicht der Mangel an hochentwickelten Baumethoden die eigentlichen Hindernisse zu einer Verbesserung der Lage sind. Das gleiche gilt aber auch für Europa. Neue Bauwerke richten sich nach den Regeln der Spekulanten und nie nach geplanten und erprobten Konzeptionen. In Spanien habe ich am Entwurf von Schulen teilgenommen, die auf Abfallgruben oder in der Umgebung von Irrenhäusern gebaut werden mußten, weil der Gemeinde kein anderer Boden zur Verfügung stand.

In Frankfurt leitete ich einen Bau, der dank der geschichtlichen Vergangenheit derart zwischen zwei Gebäuden eingekeilt war, daß wir keinen Kran, sondern einen Lift errichten mußten, um unser Material zu transportieren. Die Ironie dieser Erfahrung liegt darin, daß die ganze Gegend nach dem Krieg eine grenzenlose Schutthalde war.

Die Bauindustrie kann sich nur erfolgreich durchsetzen, wenn Boden und Infrastrukturen zur Verfügung stehen. Die Regierungen müssen entweder die notwendigen Maßnahmen ergreifen, um Boden freizumachen, oder offen gestehen, daß sie für die Probleme der menschlichen Umwelt keine Interessen haben.

has attempted to encourage the building industry to adopt and experiment with new building methods. Nobody can deny that the American building industry may use some fresh ideas, but to assume that the problem of the American slums, or of the sprawling lifeless suburbs can be solved by new techniques, is fanciful.

It is very clear to everyone that only the radicated interests of the landowners are the real obstacles to any improvement in the situation; not the lack of an advanced building technique. But the same problem is common in Europe.

New dwellings are built according to the rules of speculation, never according to planned and justified conceptions.

In Spain I have seen, and participated in the design of schools built on garbage dumps or next to mental hospitals, because they were the only land available to the municipality.

In Germany, in Frankfurt, I had to run a building site that because of past history was wedged between other buildings to the point that we could not use a crane and we had to utilize a lift for moving the materials. The irony of the experience lies in the fact that the entire area had been a waste of undistinguishable rubbish after the war.

The building industry can only operate successfully when land and the relative infrastructures are available. The governments must either take the necessary measures to insure this availability or frankly admit that they are not interested in the problem of the human environment.

7.3 Architektonische Entwicklungen

Ich möchte unter diesem ziemlich allgemeinen Untertitel die Entwicklungen beschreiben, die sich in der Morphologie des Bauwerkes, in der Methodologie des Entwurfs und im Informationsbereich abspielen werden.

Gebäude zeigen heute – gleichgültig ob industrialisiert oder konventionell – dieselbe allgemeine Konzeption wie vor Jahrhunderten. Kleinere Änderungen wurden in den mechanischen Subsystemen vorgenommen, wie etwa mit der Einführung von Klimaanlagen; doch wäre ein Urgroßvater eigentlich nicht schockiert, wenn er uns heute zu Hause besuchen würde.

Gerade das industrialisierte Bauen befindet sich in einem absurden Zustand. Praktisch versuchen wir immer noch, eine Kutsche ohne Pferd zu produzieren; wir haben den Schritt zum Auto noch nicht vollzogen.

7.3 Architectural developments

Under this very general sub-title I would like to cover the developments which will take place in the morphology of the building, in the design methodology and in the information field.

Buildings today – industrialized or not – still show the same general conception of buildings hundreds of years ago. Minor changes have been introduced in the mechanical sub-systems, with the introduction of air-conditioning, but in general no great-grandfather would suffer from shock by joining us in our dwellings.

Particularly with industrialized building the situation is rather absurd. We are still in reality attempting to produce a horse-carriage without horses, and we have not yet decided to produce a motor car. The problem is not simply a formal one but creates deep-reaching

Ashorne Hill Residential Management Training College. Architekten: Boscoe und Stanton, London. Ein «modernes» Bauwerk, aber die Konnotation unterscheidet sich kaum vom alten.

Ashorne Hill Residential Management Training College. Architects: Boscoe and Stanton, London. A "modern" building, but the general connotation remains the same as in the old ones.

Das Problem liegt nicht nur im Formalen, sondern bereitet tiefgreifende technische Schwierigkeiten. Wir wollen nun, ohne uns in komplexe Einzelheiten zu verlieren, einen grundlegenden Aspekt eines Bauwerkes betrachten: seine erwartete Lebensdauer.

Man erwartet heute von einem Bauwerk eine Lebensdauer von mindestens fünfzig Jahren. Doch können auch befriedigende Gebäude mit kürzerer Lebenszeit hergestellt werden, wobei dann aber verschiedene Hindernisse auftauchen: die Finanzierung bietet Schwierigkeiten, Versicherungen schnellen in die Höhe (warum eigentlich?) und so fort. Jetzt wissen wir aber ganz genau, daß unserer gegenwärtigen Gesellschaft schnelle und auch unerwartete Änderungen widerfahren. Warum soll man also allzu teure Bauten erstellen, die vielleicht in einem Bruchteil ihrer Lebenserwartung funktional veraltet sind, wenn wir Gebäude zu einem Zehntel dieser Kosten mit einer voraussichtlichen Lebensdauer von zehn Jahren bauen können?

Doch auch von einem rein formalen Standpunkt aus betrachtet nehmen sich die meisten industrialisierten Bauwerke irgendwie absurd aus. Auch in einem gelungenen Entwurf wie bei der Universität von York ist der Wunsch eindeutig erkennbar, die industrialisierten Komponenten dergestalt zu entwerfen und einzusetzen, daß «der Eindruck eines konventionellen Bauwerkes entsteht».

technical difficulties. Without loosing ourselves in complex details let us consider a major aspect of a building: its expected life-span.

Everybody today expects a building to last at least fifty years. However, satisfactory dwellings can be produced with a shorter life span, but then all kind of difficulties arise: financing becomes difficult, insurance more expensive (why?), etc. Now we know quite well that our present society is undergoing rapid and — to a certain extent — unforeseeable changes. Why, then, build very expensive buildings which may become functionally obsolete in a fraction of their life expectancy, when at a tenth of that cost we can produce buildings with a planned life-span of ten years?

But also from a purely formal angle most of our industrialized buildings look slightly absurd. Even in a good design like the University of York, the desire is evident to use the industrialized components (and to design them) in such a way as "to make the building look like a conventional one".

Particularly distressing are cases like the mobile homes (the first industrialized product which has attempted to break away from the conventional building concept) where to a new conception of building all kind of trappings are added (fake brick walls, iron doors, etc.) to recreate the image of the "classic" dwelling.

Several studies seem to demonstrate that people find it hard to change their attitudes towards their dwell-

Universität von York, England. Architekten: Robert Mathew, Johnson, Marshall.

University of York, England. Architects: Robert Mathew, Johnson, Marshall.

Kirche «Holy Rood Bagworth». Architekten: Blockley, Goodwin und Warner. Konnotation und Denotation werden nur oberflächlich vom Gebrauch industrialisierter Baumethoden berührt.

The Church of the Holy Rood Bagworth. Architects: Blockley, Goodwin and Warner. Connotation and denotation of this building are only superficially touched by the utilization of industrialized building methods.

Besonders gravierend sind solche Fälle wie die «Mobile Homes», das erste industrialisierte Produkt, das sich vom konventionellen Baukonzept zu lösen suchte, aber bei dem sich zu einer neuen Baukonzeption noch allerlei Putz gesellte (Backsteinimitationen, beschlagene Türen usw.), um das Bild eines «klassischen» Wohnbaus aufrechtzuerhalten.

Aus diversen Untersuchungen geht anscheinend hervor, daß es den Leuten schwerfällt, ihre Einstellung zu ihren Behausungen zu ändern, und daß sie sich Neuerungen widersetzen. Ich persönlich zweifle sehr an der Stichhaltigkeit dieser Aussagen. Vielleicht geht es einfach darum, daß ein «modernes» Bauwerk entweder gar nicht verfügbar ist oder, wenn erhältlich, der Kaufpreis zu hoch ist.

Wir müssen zu den rein formalen Aspekten die bereits erwähnten neuen Anforderungen (7.1) bezüglich der zeitlichen Flexibilität hinzurechnen, die wahrscheinlich die Schließung der Lücke zwischen der Technologie und dem Bild des Produktes erzwingen könnten.

Die Anwendung und der bereits ausgewiesene Erfolg des industrialisierten Bauens haben die Notwendigkeit eines korrekten Entwurfes aufgezeigt: von der tatsächlichen Form des Bauwerkes bis zur letzten Schraube und Bolzenverbindung.

Diese Notwendigkeit hat ihrerseits wieder gezeigt, daß unserer Entwurfsmethodologie die Präzision und die Kontrollsysteme fehlen, die eine zuverlässige Quelle für industrialisierte Produkte gewährleisten könnten. Es geht hier wiederum um den quantitativen Faktor: ein Fehler in einem einzelnen Bau wird die Kosten des

ings and tend to resist changes. Personally I seriously doubt the accuracy of these statements. Possibly the point is simply that a "modern" building is not available or that, when available, the purchase price is too high.

To the merely formal aspects we have to add the already-mentioned new requirements (7.1) for flexibility in time, which may well force a closing of the gap between the technology used and the image of the product.

The introduction of, and the success already achieved by the industrialization of building, have shown the necessity for a correct design: from the actual form of the building to the last screw and bolt connection. This requirement has shown in its turn that our design methodology lacks the precision and the checking systems to be a reliable source for industrialized products.

(Again the quantity-factor: a mistake in a single building will not greatly increase the cost of the product and/or may effect a limited amount of people. A mistake in an industrialized product may make life unbearable for a few thousands and may represent the difference between success and failure.) Considerable efforts have been undertaken at a theoretical level to provide the designer with a reliable and suitable methodology.

The works of G. Ciribini in Italy, of H. Rittel in Germany, of C. Alexander in the United States, while following different paths, show great possibilities. Unfortunately none of these methodologies have found their way

Produktes nicht wesentlich erhöhen und/oder eine begrenzte Zahl von Leuten treffen. Ein Fehler in einem industrialisierten Produkt kann aber ohne weiteres das Leben einiger tausend Bewohner unerträglich machen und den Unterschied zwischen Erfolg und Mißerfolg bedeuten. Es sind auf theoretischer Ebene beträchtliche Anstrengungen im Gange, um den Entwerfer mit einer zuverlässigen und adäquaten Methodologie zu wappnen.
In den Werken von G. Ciribini in Italien, von H. Rittel in Deutschland, von C. Alexander in den USA zeichnen sich vielversprechende Möglichkeiten ab, auch wenn sie verschiedene Richtungen einschlagen. Da leider keine dieser Methodologien in massiver Form den Weg in die tägliche architektonische Praxis gefunden hat, ist ein «Feedback» immer noch ausgeblieben.
Eine zuverlässige Entwurfsmethodologie, die sich auf allen Ebenen bewährt, ist wohl ein grundlegendes Ziel der zukünftigen architektonischen Entwicklung.
Ein drittes Ziel, das mit den vorhergehenden eng zusammenhängt, ist die Entwicklung einer organischen Speicherung und Wiederauffindung von Informationen. Das Baugewerbe leidet hier mit anderen Industrien an der unvorstellbaren Armut des gegenwärtigen Informationssystems. Täglich erscheinen relevante Dokumente, um dann im Morast von riesigen Speichern vorhandener, aber unauffindbarer Informationen zu versinken. Täglich versuchen Entwerfer und jene, die am Entscheidungsprozeß beteiligt sind, verzweifelt, Informationen aufzufinden. Das Problem wird natürlich noch durch Sprachbarrieren vertieft; ein tüchtiges

in significant quantity into the daily architectural practice, and therefore a feedback is still unavailable.
A reliable design methodology, utilizable at all levels, can be considered as a major objective for our future architectural development.
A third development, closely connected with the preceding one is the development of an organic system of information storage and retrieval. Here the building trade joins the other industries in the unconceivable poverty of our present information system. Daily relevant documents are produced, only to disappear in the morass of huge stores of available but unretrievable information.
Daily, designers and decision makers of all kinds try desperately to find information. The problem is clearly complicated by the language barrier, but a good information system should be capable of overcoming this difficulty.
The major cause of the present chaotic situation is simply the fact that during the last twenty years we have produced more documents than in all the past history of mankind, while our classification and manipulation of the produced information is still the one of the nineteenth century.
A certain amount of jealousy between the existing organizations has not helped to solve the problem, but more serious is the lack of a satisfactory intellectual basis for a complete information system: too many efforts have done little more than achieve a slightly improved form of decimal classification, without giving

Informationssystem sollte aber diese Schwierigkeiten überbrücken können.

Der Hauptgrund der momentanen chaotischen Lage liegt in der einfachen Tatsache, daß in den letzten zwanzig Jahren mehr Dokumente erschienen sind als schon in der ganzen Menschheitsgeschichte, während unsere Klassifikation und Verarbeitung der produzierten Informationen noch den Stand des neunzehnten Jahrhunderts aufweisen.

Die bestehenden Organisationen erschweren manchmal die Lösung dieses Problems durch ihren gegenseitigen Neid; schlimmer ist aber der Mangel einer ausreichenden intellektuellen Basis für ein vollständiges Informationssystem: zu viele Anstrengungen haben nur eine leicht verbesserte Form der dezimalen Klassifikation erreicht, ohne den Mechanismus der Wiederauffindung, Korrelation und Analyse von Informationen in Betracht zu ziehen. Auch hier werden beträchtliche Anstrengungen unternommen; sie sind aber leider über die ganze Welt verzettelt, während ein wirklich brauchbares organisches Informationssystem zum vornherein das ganze Feld beherrschen müßte, also sich über nationale und institutionale Barrieren hinwegsetzen müßte.

Der Mangel eines organischen Informationssystems könnte sehr wohl eine Barriere bilden, die jegliche Weiterentwicklung inner- und außerhalb des Baugewerbes unterbinden könnte.

Die Schwierigkeit kann auf zwei Arten entstehen. Einerseits mag es auf Grund rein mechanischer Schwierigkeiten (zum Beispiel Wiederauffinden und Korrelation von Informationen) äußerst aufwendig sein, ein spezifisches Problem zu überwinden, das – hätte man Informationen zur Verfügung – mit Leichtigkeit gelöst werden könnte.

Andererseits sehen wir bereits die enormen Schwierigkeiten der Experten, Erfahrungen auszutauschen. Die Spezialisierung der Sprache jedes einzelnen Forschers, die oft auf einer ad hoc gebildeten Terminologie beruht, errichtet allmählich neue Barrieren zwischen den Leuten und Fachgebieten.

So zeichnet sich eine Situation ab, die eine direkte Kommunikation und Assimilation von Ergebnissen verunmöglicht und nach einem speziellen «Übersetzungsfach» verlangt. Sogar im Baugewerbe, einem Gebiet, in dem wir gewiß nicht an einem Überfluß an theoretischen Grundlagen leiden, ist es mir in den letzten Jahren aufgefallen, wie unsere Sprache immer unverständlicher wird, was allen Mitwirkenden im Weg steht und die Kreativität herabsetzt.

Vom operationellen Standpunkt aus sollte es leicht einzusehen sein, daß die Priorität im Aufbau eines brauchbaren Informationssystems mit dem gesamten dazugehörigen «software» und «hardware» liegen müßte. Ebenso wäre von einem allgemeinen wirtschaftlichen Standpunkt aus die Entwicklung eines Informationssystems von Nutzen: der Kräfteverschleiß in der heutigen Forschung und Entwicklung, der durch den Mangel an Informationen verursacht wird, ist äußerst teuer und niederschmetternd.

consideration to the mechanics of information retrieval, correlation and analysis.

Here again considerable efforts are under way, but unfortunately they are world widely dispersed, while, by definition, an organic information system has to cover the whole field, beyond national or institutional barriers, to be really useful.

The lack of an organic information system may well create a barrier preventing any further development (inside and outside the building field).

The obstacle can arise in two ways. At one end, due to purely mechanical difficulties (how to retrieve, how to correlate information) we may find it extremely labourious to overcome a specific problem which, with the availability of the information, could be solved with ease.

On the other hand we already see the enormous difficulties experts have in communicating each other's experiences.

The specialization of each researcher's language (often based on ad-hoc created terminology) is gradually building up a new barrier between people and disciplines.

A situation may soon arise and possibly it already exists, where the direct communication and assimilation of findings may not be possible, and a special "translation" discipline may be required. Even in the building trade (a field in which we certainly do not suffer from an excess of theoretical material) I have witnessed during the last few years how our language has become increasingly unitelligible with the resulting frustration for all participants and the resulting decrease in creativity.

From an operational angle it would seem that our first priority should be the organization of a working information system, with all the related software and hardware.

Also from a general economical angle the development of an information system seems to make good sense: the waste of effort in research and development we have today – caused by the lack of information – is very expensive and quite staggering.

Das von Matti Suuronen, Finnland, entworfene «Futuro»-Haus ist ein Beispiel integrierter Vorfabrikation.

An example of integral prefabrication: the "Futuro"-house, designed by the architect Matti Suuronen, Finland.

8. Schlußfolgerungen

8.1 Die Zukunft des industrialisierten Bauens

Konrad Wachsmann schrieb vor mehr als zehn Jahren das Werk «Der Wendepunkt im Bauen». Wir müssen uns heute darüber klar sein: der Wendepunkt liegt hinter uns. Die andauernden Diskussionen über den Wert der Bauindustrialisierung als Lösung unserer dringenden Probleme sind obsolet und sinnlos. Dank der gegenwärtigen sozialen Lage und der demographischen Explosion liegt der einzige Ausweg nur noch in der Bauindustrialisierung. Auch andere Lösungen kommen selbstverständlich in Frage; sie liegen aber außerhalb der Macht derjenigen, die sich der Bauindustrie widmen (zum Beispiel Kriege, Massenumsiedlungen, Naturkatastrophen usw.).

Die heutige Lage in den kommunistischen wie in den kapitalistischen Ländern verlangt nach der Industrialisierung des Bauens. Die Gründe dafür sind einleuchtend und ergeben sich von selbst: dringende Wohnungsnot und großer Mangel anderer lebenswichtiger Bauten; das Verschwinden des traditionellen Handwerkers; die äußerst notwendige Verbesserung der Umweltbedingungen. Das alles ist bereits seit fünfzehn Jahren bekannt. Untersuchungen von den Vereinigten Nationen und von Einzelpersonen haben schon lange darauf hingewiesen. Wichtig ist heute die Tatsache, daß die Bauindustrialisierung nicht mehr eine wünschenswerte Neuerung, sondern ein grundlegendes Phänomen unserer sozialen Struktur bildet.

In Großbritannien werden heute über siebzig Prozent aller Schulbauten mit industrialisierten Methoden errichtet.

Die italienische Regierung ist bereit, umfangreiche Mittel für die Industrialisierung des Wohnungsbaus aufzuwenden.

Russische Wohnungen werden praktisch durchwegs mit industrialisierten Methoden gebaut.

Die USA schlagen denselben Weg ein (wobei «Operation Breakthrough» lediglich den Anfang bildet), um die untragbaren Zustände in den Städten zu lindern. Noch bedeutsamer ist die Tatsache, daß beinahe alle Staaten Europas bereits über durchschlagskräftige und straff industrialisierte Bauorganisationen verfügen, die sich voll mit diesen Problemen beschäftigen. Heute ist die französische Regierung in der Lage, einen Wettbewerb für den Bau einer neuen Stadt auszuschreiben; es ist nämlich selbstverständlich, daß genügend viele industrialisierte Organisationen daran teilnehmen werden. Es ist sogar für jeden konventionellen Unterneh-

8. Conclusions

8.1 The future of industrialized building

Over ten years ago Konrad Wachsmann wrote "The Turning Point of Building". Today we should well be clear about it: the turning point is behind us. Somehow the continuing discussions about the validity of industrialized building as a solution to our urgent problem are obsolete and futile. With the present social environment and birth-rate explosion we have to accept the industrialization of building as the only alternative. Clearly other solutions can be considered, but they are outside the control of the people involved in the building industry (through wars, resettlement policies, big natural catastrophes, etc.).

As the situation stands today, both in capitalistic and in communist countries, building has to be industrialized.

The reasons for it should be self-evident: an urgent need for housing and other living facilities; the disappearance of the traditional tradesman; a great necessity for a higher standard of environmental conditions. However, what I have already said has been well known for some fifteen years.

Studies by the United Nations and by isolated individuals have been indicating the way for a long time.

What is new today is the fact that the industrialization of building is no longer merely a desirable development but a firmly-grounded phenomenon of our social structure.

In the United Kingdom today over seventy per cent of all school buildings produced are erected by industrialized methods.

The Italian Government is prepared to invest huge amounts of money in the industrialization of housing. Practically all Russian dwellings are produced by industrialized methods.

The U.S.A. is following in the same tracks (and Operation Breakthrough is only the beginning) in order to alleviate the intolerable conditions in urban environments.

What may be of more practical relevance is the fact that almost all European countries already have strong and well-structured industrialized building organizations, coping with the problem on a routine basis.

Today the French Government is able to open a competition for the construction of a new town in the quiet assurance that sufficient industrialized organizations will participate. Even more, it is well-nigh impossible for any conventional contractor or architect

mer oder Architekten praktisch unmöglich, an den oben erwähnten Wettbewerben mit Erfolgsaussichten teilzunehmen. Sie können der Macht der stärkeren industrialisierten Organisationen nicht entgegenwirken. Man darf mit guten Gründen behaupten, daß in einem Wettbewerb für alle größeren Bauprojekte nur industrialisierte Organisationen auf Erfolg rechnen können. Sogar in Spanien wurde 1970 ein Wettbewerb für annähernd 4500 Wohnungen mit Bedingungen ausgeschrieben (Bauzeit achtzehn Monate), an denen konventionelle Bauunternehmer zum vornherein scheitern mußten.

Wir haben den Punkt erreicht, wo die Aussichten der konventionellen Bauweise (als Architekt oder Unternehmer), einen größeren Auftrag zu erhalten, genau so klein sind wie etwa bei einem alleinstehenden Ingenieur, der sich um die Konstruktion eines Überschallflugzeuges bewerben möchte.

Im Lichte der gegenwärtigen Marktsituation müssen wir uns jedoch noch eine andere Tatsache vor Augen halten: die unzähligen kleinen Unternehmer, Architekten und Ingenieure, die noch immer erfolgreich tätig sind oder wenigstens als Nachhut den Kampf in der Verteidigung der konventionellen Bauweise weiterführen.

Eine Frage hat die Bauindustrialisierung außer acht gelassen: wie wird sich wohl die Zukunft dieser Organisationen und Fachleute gestalten? (Ich habe bereits in Kapitel 2 erwähnt, daß die Arbeiter vom Industrialisierungsprozeß kaum berührt werden.) Was geschieht mit dem Talent und dem fachmännischen Geschick, das in diesen kleinen Organisationen und Betrieben steckt? Vordergründig könnte man antworten, daß ihre Tage wahrscheinlich gezählt sind und daß einzelne Personen von den größeren industrialisierten Organisationen aufgesogen werden. Oder aber sie könnten sich zusammentun und auf der Schaubühne als neue, größere und lebensfähigere Organisationen auftreten.

Das Problem liegt meines Erachtens eigentlich nicht in der Frage, was mit diesen Organisationen und Betrieben geschehen soll, sondern eher darin, welchen Einfluß ihr Verschwinden auf die freie Wahl des Benutzers und auf die Qualität der Umwelt haben wird.

Schließlich hatten kleine Bauunternehmer, Architekten und Ingenieure Zeit genug, die Schrift an der Wand zu lesen, und sie haben inzwischen lange genug von den günstigen Umständen profitiert. Beinahe die gesamte verunstaltete städtische Umwelt, die Verschandelung der Küsten, Bergkurorte und der Landschaft sind ihrer Tätigkeit zu verdanken. Es wäre verfehlt, sich allzusehr um ihre Zukunft zu kümmern. Wichtiger ist, ob die industrialisierten Bauorganisationen fähig sind, sich besser als ihre Vorläufer zu bewähren.

Wenn wir einen Vergleich mit andern Industrien ziehen, fällt die Antwort zweifellos negativ aus, wenigstens was den qualitativen Aspekt betrifft. Alle Industriezweige haben ihre Fähigkeit bewiesen, riesige Quantitäten von Produkten herzustellen (sei es Kriegsausrüstung, Pharmazeutika, Anzüge oder Automobile), aber recht wenige haben sich um die Qualität ihres Produktes gekümmert. Die Dichotomie «Qualität-Quantität» ist heute das Hauptproblem des industrialisierten Bauens.

Dazu ist das Qualitätsproblem bei Bauwerken viel wichtiger als bei andern Produkten. Wir verbringen

to participate successfully in the above-mentioned competitions. Stronger industrialized organizations will simply crush their efforts under their own weight. We can safely state that for all larger building projects only industrialized organizations can hope to compete successfully. Even in Spain, in 1970, a competition for approximately 4500 flats was organized under such conditions (erection time eighteen months) as to make the participation by conventional builders impossible.

We have reached the point where to win a large contract using conventional methods (as an architect or a contractor) is as unlikely as it would be for an isolated engineer to obtain a contract for the construction of a supersonic airliner.

However, against the reality of the present market situation we have to consider another reality: that of the thousands of small contractors, architects and engineering practices which are still successfully operating, or are at least continuing to fight a rearguard battle in the defence of the conventional method of building.

One of the questions which nobody has cared to answer is what might be the future of these organizations and professionals?

(I have already mentioned in chapter 2 that the workers are hardly affected by the process of industrialization.)

What will become of the talent and expertise available in these small organizations and practices? The easy answer is that they will slowly die out, individuals being absorbed into the bigger industrialized organizations. Another possible answer is that they could merge together and appear on the scene as bigger, more viable organizations.

In my opinion the problem is not really that of what should happen to these organizations and practices, but rather what influence their disappearance will have upon the freedom of choice of the user and upon the quality of the environment.

Finally small contracting-organizations, architects and engineering practices have had time enough for reading the writing on the wall, and in the meantime have exploited long enough a set of favourable circumstances.

Most of the appalling urban environments, most of the rape perpetuated on our beaches, mountain resorts and countryside are the results of their activity.

To be too much concerned about their future seems to me rather uncalled for. The real question lies in whether the industrialized building organizations are capable of doing better than its predecessors.

If we have to make a comparison with other industries, the answer is certainly negative, at least from a qualitative point of view. All branches of industry have proved their ability to provide large quantities of products (be they war-equipment, pharmaceuticals, suits or motor cars) but very few indeed have shown any interest in the quality of their product. The dicotomy "quality-quantity" is the main problem today for the industrialized building.

Moreover, the problem of quality is of more importance for buildings than for other products: we spend most of our time in buildings or admidst them and only a fraction of it in cars or aircrafts (they also last longer,

einen großen Teil unserer Zeit in oder um Bauten und nur einen Bruchteil davon in Autos oder Flugzeugen (sie haben auch ungeachtet der Qualität eine längere Lebensdauer). Bauwerke üben einen tiefgreifenden Einfluß auf unsere Existenz aus, und zwar auf den Einzelnen wie auf die Gesellschaft. Es ist bemerkenswert, daß die kapitalistischen wie die kommunistischen Regierungssysteme bis jetzt drastisch versagt haben, adäquate Lebensbedingungen zu schaffen. Sie geben sich gewöhnlich mit Ausreden wie Mangel an finanziellen Mitteln, Notlage und ähnlichem zufrieden.

Die heutige Lage beweist, daß die Bauindustrialisierung wenigstens über die potentielle Leistungskraft verfügt, adäquate Lösungen zu finden, während die konventionelle Bauweise nur noch zu chaotischeren Zuständen führt. Die Gegenwart und Zukunft werden also unweigerlich von der Industrialisierung des Bauens bestimmt, und wir müssen endlich die unrealistische Alternative der konventionellen Bauweise verwerfen.

Konventionelle Arbeitsmethoden können in einer reichen Gesellschaft noch immer eine Rolle spielen und für Prestigebauten eingesetzt werden, womit sich Direktoren, Verwaltungsräte, erfolgreiche Schauspieler usw. schmeicheln können. Für die überwältigende Mehrheit sind aber nur industrialisierte Bauten relevant. Das Problem ist also, wie wir eine angemessene Qualität garantieren können, damit wir nicht zu Gefangenen unserer eigenen Technologie werden.

with or without quality). Buildings affect deeply our way of life both as individuals or as a society. It may be interesting to note that until now both government systems, the Capitalistic and the Communist, have failed dramatically to create adequate living environments (usually utilizing the spurious arguments of unavailable financial resources, emergency situations and the like).

As the situation stands today we know that industrialized building does at least have the potential capacity to provide an adequate solution, while conventional building can only add to the existing chaos. We must therefore accept that our present and our future will be determined by the industrialized building and dispose once and for all of the non-existant alternative of the conventional building methods.

Conventional techniques may continue to play a rôle in a rich society by providing prestige-buildings and satisfying the ego of company presidents, boards of directors, successful actors and so on. For the vast majority of the population only industrialized buildings are of any relevance.

The problem is, how can we ensure satisfactory quality and avoid becoming prisoners of our own technology?

8.2 Die Notwendigkeit eines Kontrollorgans

Die vorausgehenden Kapitel und die Beurteilung der Lage, die sich jedem aufzwingt, der unsere heutigen Städte betrachtet, unterstreichen zwei Aspekte des gesamten Problems:

a) Die konventionelle Bauweise ist nicht imstande gewesen, auf unsere Bedürfnisse einzugehen.

b) Bis heute sind Kontrollorgane, seien es Architekten, Ortsbehörden, Regierungen oder individuelle Spekulanten, unfähig, die Interessen der Gemeinschaft wahrzunehmen.

Wir wissen, daß die Bauindustrialisierung ohne durchschlagskräftige und geeignete Kontrollorgane unsere Umwelt negativ beeinflußen wird. Sich auf den selbst kontrollierenden Mechanismus der Privatindustrie («Wir müssen gute Produkte herstellen, sonst kaufen sie die Leute nicht mehr») oder die aufgeklärten Selbstinteressen der Politiker («Wenn wir uns nicht für einen adäquaten Wohnungsbau einsetzen, werden wir nicht mehr gewählt») zu verlassen, wäre unglaublich naiv und hätte – was unsere Umwelt betrifft – beinahe etwas Selbstmörderisches an sich.

Die Benutzer selber müssen, mit der Unterstützung von hochqualifizierten Fachleuten, eine direkte Kontrolle über die Bautätigkeit ausüben. Dieses Vorgehen wird selbstverständlich mit den üblichen technischen Argumenten angezweifelt: «Das Problem ist zu komplex, als daß man es in die Hände des Volkes legen könnte», «das Volk wird nie den Umfang der vielen Probleme begreifen können», «die Leute sind desinteressiert» und so fort.

8.2 The need for a controlling authority

The preceding chapters, including the direct assessment each one of us can easily make by looking at our present urban environment, presents us with two factors of the general problem:

a) Conventional building methods have failed in providing a satisfactory answer to our needs.

b) Up to today controlling-agencies, be they architects, the local authority, the central government, or private speculators, have not been capable of safeguarding the interests of the community.

We know that the industrialization of building will only contribute towards the degradation of our environment if effective and proper forms of control are not established. To believe in the self-controlling mechanism of the private industry ("we must produce good products or the public will not buy them") or the enlightened self-interest of the public authority ("if we dont give them good housing they will remove us from power") would be incredibly naïve and in the case of our living environment almost suicidal.

The users themselves, with the help of highly-controlled professionals, must create some form of direct control over the building activity.

Clearly this kind of statement will be disputed by the usual technical arguments: "The problem is too complex to be left in the hands of the people", "the people will never be capable of understanding the full implications of all the problems", "people are not interested", and so on.

Some of these statements are indeed correct, today.

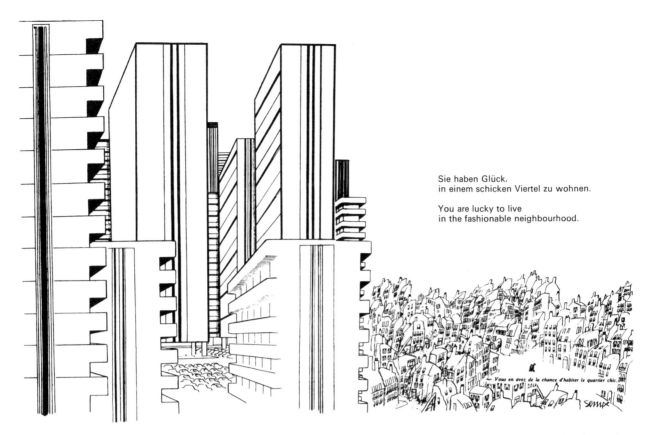

Sempés Zeichnung ist ein ausgezeichneter Ausdruck des Versagens unserer heutigen Bauprogramme.

This drawing by Sempé is an excellent expression of the failure of our present building programmes.

Einige dieser Einwürfe haben heute tatsächlich eine Berechtigung. Die Leute müssen jede Art von Behausung, die ihnen dargeboten wird, lediglich annehmen und kaufen. Festgefahrene Barrieren bürokratischer oder einfach sprachlicher Natur hindern sie daran und entmutigen sie, irgendeine kontrollierende Funktion auszuüben.
Das heißt nun keineswegs, daß die Leute «immer» unfähig sind, an der Gestaltung und Kontrolle ihrer Umwelt mitzuwirken. Es ist eigentlich der größte Fehler der heutigen Architekten, daß sie Sprachbarrieren errichtet haben, um den Laien zu entmutigen, direkt am Entwurfsprozeß teilzunehmen, während sie sich scheinheilig als «Verteidiger der Interessen des Volkes» ausgeben.
Um das vorhandene Potential der Bauindustrialisierung voll auszuschöpfen, müssen wir die Bewohner in den Prozeß einspannen. Und dieses Ziel kann erreicht werden. Es können unabhängige Organisationen geschaffen werden, die sich den Anforderungen der Benutzer und ihrer wahren Bedürfnisse widmen. Unabhängige Aufsichtsorgane aus Fachleuten und Laien können eingreifen (in den meisten Ländern selbst im Rahmen der gegenwärtigen Gesetzgebung), um unzulässige Bauprojekte zu Fall zu bringen. Politische Parteien und andere öffentliche Institutionen können auf Bauprobleme aufmerksam gemacht werden und mit gezielter Unterstützung einen sehr großen kontrollierenden Einfluß bereits in den Anfangsstadien der Planung ausüben. Um aber dieses Ziel zu verwirklichen, brauchen wir «Promoter». Für mich ist natürlich der Architekt derjenige, der aktiv mitmachen sollte bei der Aufstellung solcher Organisationen, die das Bewußtsein der Leute wachrütteln und ihnen bei

People are obliged to accept and buy any kind of dwelling which is offered to them. Well-established barriers of bureaucratic or simply of language nature, prevent them and discourage them from taking part in any kind of controlling-function.
This, however, does not mean that people will "always" be incapable of participating in the design and in the control of their environment. It is practically the worst failure of the today architects that they have established language barriers to discourage the layman from a direct intervention in the design process, hypocritically assuming the position of "defenders of the people's interests".
To make a good use of the potential offered to us by the industrialization of building we have to involve the common people in the process. And this can be achieved. Independent organizations can be established to study the requirements of the users and their real needs. Independent watch-dog committees formed by professionals and laymen can intervene (even under present legislation in most countries) to prevent unsatisfactory building developments. Political parties, and other organized institutions, can be made aware of the problems of building and, properly assisted, can achieve a very meaningful degree of control participating in the early planning-stages.
However, to achieve all of this we need promoters. It is natural for myself to see in the architect the person who should help in establishing such organizations, who should awake the conscience of the common people, who should assist them in defending their present and future living-environment.
Most likely, however, the architectural profession has degenerated too much to be capable of assuming such

der Verteidigung ihres gegenwärtigen und zukünftigen Lebensraumes helfen.

Wahrscheinlich ist aber der Architektenberuf bereits zu degeneriert, um eine solch führende Rolle einzunehmen. Das Prestige dieses Faches wurde zunichte gemacht von allzu vielen Architekten, die sich «verkauft haben». Wenn diese Behauptung stimmt, hoffen wir nur, daß andere Berufe gewillt sind, die Verantwortung zu übernehmen und die notwendige Kontrolle auszubauen.

Die Bauindustrialisierung ist eine Tatsache. Sie kann uns allen adäquate Lebensbedingungen schaffen. Sie ist vielleicht auch ein neues Glied in der Kette, mit der uns die Zivilisation fesselt. Es liegt an uns, aus dieser neuen Technologie einen nützlichen Diener und nicht einen neuen Tyrannen zu machen.

a leading rôle. Its prestige has been destroyed by too many architects "selling out". If this judgement is correct we can only hope that other professions will be willing to take over the responsibility and perform the necessary duty of control.

The industrialization of building is with us. It can give us all good living conditions. It may also be a new link in the chains our civilization is imposing upon us. It is up to us to make this new technology a useful servant and not a new tyrant.

9. Verzeichnis der Begriffe   9. Glossary

Modul: Spezieller Maßwert für dimensionale Koordination (OEEC).

Grundmodul: Ein grundsätzlicher Modul mit einem festen Wert, um die Größe der Komponenten mit der größten Flexibilität und Einfachheit zu koordinieren (OEEC). (Der Grundmodul beträgt in Europa und Amerika 10 cm oder 4 Zoll.)

Modulraster: Ein geradliniges Bezugssystem, in dem die Linien in Abständen folgen, die aus dem Mehrfachen des Grundmoduls bestehen (Modular Society). (Die gebräuchlichsten horizontalen modularen Raster sind heute: 60 cm, 90 cm, 120 cm; oder 2 Fuß, 3 Fuß, 4 Fuß.)

Planungsmodul: Ein Mehrfaches des Grundmoduls, das für den Entwurf von Bauwerken verwendet wird.

Baustoff: Ein Material ohne bestimmte geometrische Form (pulverartig, körnig, faserig, flüssig oder breiartig), z.B. Asphalt, Kies, Zement, Holzfaser (OEEC).

Teil: Teilstück, aus dem sich eine größere Einheit zusammensetzt (z.B.: die Kreuzsprosse eines Fensters, das Türblatt).

Zubehör: Kleine Einheiten, die bei der Montage der Komponenten gebraucht werden (z.B. Schrauben, Nägel, Verbindungsstücke).

Bauhalbzeuge: Baustoffe, produziert als Halbzeuge, gewöhnlich durch einen kontinuierlichen Ablauf hergestellt, mit festgesetztem Querschnitt und unbestimmter Länge, gewalzte, gezogene, extrudierte und gesägte Produkte, z.B. Profile, Röhren, Leitungen, Bohlen, Wandbretter, Draht und Kabel, Bleche und Platten, gesägte Bretter usw. (OEEC).

Komponenten: Montage von Baustoffen und Teilen mit einer spezifischen Funktion oder spezifischen Funktionen (z.B. ein Fenster, ein Wandpaneel, ein Balken). (Die OEEC braucht das Wort «zusammengesetzter Bauteil» in einem ähnlichen Sinn.)

Subsystem: Ein funktional aufeinander bezogener Satz von Komponenten (andere Autoren nennen das ein «Element» oder ein «funktionales Element»), z.B. die Struktur, die Außenwand, das Dach.

Module: The common unit of measure particularly specified for dimensional coordination (O.E.E.C.).

Basic module: A fundamental module whose value is fixed to coordinate the sizes of components with the greatest flexibility and convenience (O.E.E.C.). (The basic module is 10 cm or 4 inches for most European and American countries.)

Modular grid: A rectilinear reference grid in which the lines are spaced in multiples of the basic module (Modular Society). (Most common horizontal modular grids today are: 60 cm, 90 cm, 120 cm, or 2 feet, 3 feet, 4 feet.)

Planning module: A multiple of the basic module used for the design of buildings.

Material: A building material having no definite geometrical shape (powder, granular, fibrous, liquid or paste), e.g.: asphalt, gravel, cement, wood fibre (O.E.E.C.).

Part: Any portion of which something is composed (Webster Dictionary – e.g.: the transom of a window, the leaf of a door).

Bits and pieces: Small units utilized in assembling components (e.g.: screws, nails, fasteners).

Sections: Building materials produced in semi-finished form, usually manufactured by a continuous process, of definite cross section and unspecified length, e.g.: rolled, drawn, extruded or sawn products as: sections, tubes, pipes, planks, wall-board, wire and cable, sheet and plate, sawn board, etc. (O.E.E.C.).

Component: Assembly of materials and parts with a specifically defined function or functions (e.g.: a window, a wall panel, a beam). (The O.E.E.C. utilizes the term "assembly" for a similar purpose.)

Sub-system: A functional inter-related set of components (other authors call this an "element" or a "functional element"), e.g.: the structure, the outside wall, the roof.

System: A set of dimensionally and physically inter-related sub-systems and components that form a build-

System: Ein Satz von dimensional und physikalisch aufeinander bezogenen Subsystemen und Komponenten, die ein Bauwerk bilden. (Die OEEC definiert ein «Bausystem» als gleichzeitigen Gebrauch von ausgewählten Baukomponenten, um ein Bauwerk zu errichten.)

Bauwerk: Werke, die durch die Tätigkeit des Bauens entstehen (OEEC), ... um eine Funktion spezifischer sozialer Organisationen zu erfüllen (van Bogaert). Bemerkung: Wir können unter gewissen Bedingungen die Ausdrücke «System» und «Bau» synonym gebrauchen und ein Bauwerk als das Ergebnis der Kombination einer bestimmten Anzahl von Subsystemen definieren. Danach können alle Bauwerke (ob mit konventionellen oder industrialisierten Methoden errichtet) als Systeme bezeichnet werden; wir müssen dann den Begriff «modulares Bausystem» für jene Systeme gebrauchen, die unter den Regeln der modularen Disziplin stehen.

Componenting: Ein organisatorischer Vorgang, durch den verschiedene unabhängige Hersteller Komponenten oder Subsysteme auf den offenen Markt bringen, die dimensional und physikalisch in ein System integriert werden können. Bemerkung: Ein solches System wäre selbstverständlich modular und könnte als ein «offenes System» definiert werden.

Geschlossenes System (modular): Ein Paket von dimensional und physikalisch aufeinander bezogenen Subsystemen und Komponenten, gewöhnlich unter der organisatorischen Aufsicht einer einzelnen Organisation (z. B. CLASP, SEAC, Constructions Modulaires).

Örtliche Vorfertigung: Die Produktion von Komponenten auf der Baustelle, die für die Konstruktion eines Bauwerkes verwendet werden.

Baumethode: Eine Kombination von Techniken, die bei der Konstruktion von Bauwerken gebraucht werden.

Bemerkung: Für eine vollständige Terminologie wird der Leser auf die Werke in der Bibliographie verwiesen.

ing. (The O.E.E.C. defines a "System of building" as: simultaneous use of selected building components to form a building.)

Building: Works resulting from the act of construction (O.E.E.C.),... to fulfill a function of specific social organization (van Bogaert).
Remark: We may use the term "system" as synonymous of "building" under certain conditions, and define a building as the result of the additions of a set of sub-systems. In doing so all building (erected by conventional or industrialized methods) can be described as systems, and we have then to use the term "modular building system" for the systems operating under the rules of the modular discipline.

Componenting: An organizational approach by which several independent manufacturers offer on the open market components or sub-systems which can, dimensionally and physically, be integrated in a system.
Remark: Such a system would obviously be modular and could be defined as an "open system".

Closed system (modular): A package of dimensionally and physically inter-related sub-systems and components, usually under the organizational control of a single agency (e.g.: CLASP, SEAC, Constructions Modulaires).

Site-prefabrication: The production on the building site of components to be used for construction of the building.

Methods of building: A combination of techniques used to erect a building.

Remark: For a complete terminology the reader is referred to the documents mentioned in the bibliography.

10. Bibliographie/Bibliography

Diamant, R.M.E.: Industrialized Building 3, London, 1968

Meyer-Bohe, Walter: Vorfertigungs-Atlas der Systeme. Vulkan Verlag, Essen, 1964

Nouaille, R.: La préfabrication. Eyrolles Editeur, Paris, 1957

— : Industrialisierung im Wohnungsbau, Literatur, Dokumentation, Studien-Gemeinschaft für Fertigbau, März, 1967

O.E.E.C.: Modular Coordination in Building. Paris, 1961

O.E.E.C.: Prefabricated Building, Paris, 1958

E.F.L.: School Construction Systems Development. *An Interim Report.* Educational Facilities Laboratories, New York, N.Y., November, 1965

E.F.L.: School Construction Systems Development. *S.C.S.D.: The project and the schools.* Educational Facilities Laboratories, New York, N.Y., 1967

E.F.L.: Study of Educational Facilities. *Report T-1: Introduction to the first SEF Building System.* Toronto: The Metropolitan Toronto School Board Study of Educational Facilities, March, 1968

C.I.B.: Towards Industrialised Buildings, Elsevier Publishing Co., New York, Amsterdam, 1967

Kelly, Burnham: The Prefabrication of Houses, M.I.T., Cambridge, 1967

Wachsmann, Konrad: The Turning Point of Building. Reinhold Publishing Co., New York, 1967

White, R.: Prefabrication, A History of Its Development in Great Britain. H.M. Stationery Office, London, 1965

E.C.E.: Proceedings of the Seminar on Changes in the Structure of the Building Industry, United Nations, Geneva

Bussat, Pierre: Die Modul-Ordnung im Hochbau, K. Krämer, Stuttgart, 1963

U.N.: Modular Coordination in Building. United Nations, New York, 1966

Ehrenkrantz, Ezra: The Modular Number Pattern. Alec Tiranti, Ltd., London, 1956

Koncz, Tihamer: Handbuch der Fertigteilbauweise. Bauverlag, Berlin, 1967

Von Halasz, R.: Industrialisierung der Bautechnik, Werner Verlag, Düsseldorf, 1966

Beyer, Glenn H.: Marketing Handbook for the Prefabricated Housing Industry, Cornell University, Ithaca, New York, 1955

A.I.A.: Emerging Techniques of Architectural Practice. Washington, D.C., 1966

C.I.B.: Innovation in Building, Elsevier, Amsterdam, 1962

Bishop, D.: The Economics of Industrialised Building, Gaston, England, 1966

Bishop, D.: Industrialised Techniques in Housing, Moscow, 1963

Deatherage, George E.: Construction Company Organization and Management. McGraw-Hill, New York, 1966

Ministry of Education: The Story of C.L.A.S.P., HMSO, Ministry of Education Bulletin No. 19, London, 1967

Europrefab: Systems Handbook, Interbuild Publications Ltd., London, 1969

Brandle, Kurt: Bauen mit Elementen, Kombinationen; Studienhefte zum Fertigbau, Heft 7/8, Vulkan Verlag, Essen, 1968

Lorker, Eric: Modular Primer, Modular Society, London, 1963

Caporioni – Garlatti: La coordinazione modulare, Marsilio Editore, Padova, 1964

T. Schmid – C. Testa: Bauen mit Systemen, Artemis Verlag, Zürich, 1969

Fuzio, Giovanni: Industrializzazione dell'edilizia, Dedalo Libri, Bari, 1965

Lewicki, B.: Factores de la industrialización de la construcción, Instituto Eduardo Torroja, Madrid, 1965

Paraskevopoulos, Stephen C.: Architectural Research on the Structural Potential of Foam Plastics for Housing in Underdeveloped Countries, University of Michigan, Ann Arbor, 1966

Spadolini, Pierluigi: Prefabbricazione in cantiere, Libreria Editrice Fiorentina, Firenze, 1969

B.I.T.: International Labor Office, Social Aspects of Prefabrication in the Construction Industry, Geneva, 1968

Thomas, Richard K.: Three Dimensional Design. A Cellular Approach. Van Nostrand-Rheinhold Co., New York, 1969

Benček, Jan: Prefabrikacia, Polygraf, Bratislava/Preßburg, 1968

Van Bogaert, Georges: Industrialisation de la construction, Exposition EPF, Ecole Polytechnique, Lausanne, 1970

Chemetov, Paul: Création architecturale et industrialisation, Fondation pour le développement culturel, Paris, 1971

Kepes, Gyorgy: Module, Proportion, Symétrie, Rythme. La Connaissance, Bruxelles, 1968

Lewicki, Bohdan: Bâtiments d'habitation préfabriqués en éléments de grandes dimensions, Editions Eyrolles, Paris, 1965

M. K. Jeldsen – W. R. Simonsen: Industrialised Building in Denmark, A. Jespersen and Son Foundation, Kopenhagen, 1965

Döring, Wolfgang: Perspektiven einer Architektur. Suhrkamp Verlag, Frankfurt, 1970

Olivieri, G.M.: Prefabbricazione o metaprogetto edilizio, Etass Kompass, Milano, 1968

Griffin, C.W.: Systems, E.F.L., New York, 1971

Faccio, Franco: Il giunto. Associazione Italiana Prefabbricazione, Milano, 1968

Miller, David H.: Integration – The Future of Home Building. Harward University, Boston, 1959

Dietz, Albert G.: Plastic In Architecture, M.I.T., Cambridge, 1967

A.I.R.E.: Dieci studi preliminari alla industrializzazione edilizia, Milano, 1965

Ciribini, Giuseppe: Il Componenting. Edizioni E.A. Fiere, Bologna, 1968

Abraham, Pol: La casa prefabbricata. Vitali e Ghianda, Genova, 1953

Fuzio, Giovanni: Costruzioni pneumatiche. Dedalo Libri, Bari, 1968

Petrignani, Achille: Industrializzazione dell'edilizia. Dedalo Libri, Bari, 1965

A.I.T.E.C.: Industrializzazione e prefabbricazione nell'edilizia ospedaliera. Roma, 1970

Centro Studi: La prefabbricazione nell'edilizia scolastica, Ministero della Pubblica Istruzione, Roma, 1965

Royal Institute of British Architects: The Industrialisation of Building, London, 1963

B. Huber – J.C. Steinegger: Jean Prouvé. Artemis Verlag, Zürich, 1971

Frei, Otto: Zugbeanspruchte Konstruktionen (Bd. 1), Ullstein Verlag GmbH, Frankfurt, 1962

Frei, Otto: Zugbeanspruchte Konstruktionen (Bd. 2), Ullstein Verlag GmbH, Frankfurt, 1966

11. Organisationen, die sich mit dem industrialisierten Bauen beschäftigen

11. Relevant Organizations Interested in the Industrialization of Building

United Nations: Centre for Housing, Building and Planning Research, Training and Information Section, New York, U.S.A.

The Construction Research and Information Association: 6, Storey Gate, Westminster, London, S.W.1, England

National Building Agency: NBA House, Arundel Street, London, W.C.2, England

O.E.C.D.: 2, rue André-Pascal, Paris 16ᵉ, France

C.I.B.: 700, Weena, P.O. Box 299, Rotterdam, Holland

C.S.T.B.: 4, rue du Recteur-Poincaré, Paris 16ᵉ

Forest Product Laboratory: Madison, Wisconsin, U.S.A.

Building Research Station: Garston, England

National Research Council of Canada: Division of Building Research, Ottawa, Canada

The National Swedish Institute for Building Research: Varmallavägen 191, P.O. Box 27163, 10252 Stockholm 27, Sweden

Educational Facilities Laboratories: 477, Madison Avenue, New York 10022, N.Y., U.S.A.

Research Institute for Building and Architecture: Letenska 3, Prague III, Czechoslovakia

The Modular Society: London, England

Europrefab: Galleria Passarella 1, Milan, Italy